Life after New Media

Mediation as a Vital Process

Sarah Kember and Joanna Zylinska

The MIT Press
Cambridge, Massachusetts
London, England

First MIT Press paperback edition, 2015

MIT Press books may be purchased at special quantity discounts for business or sales promotional use. For information, please email special_sales@mitpress.mit.edu.

This book was set in Stone Sans and Stone Serif by Toppan Best-set Premedia Limited. Printed and bound in the United States of America.

Library of Congress Cataloging-in-Publication Data

Kember, Sarah.
Life after new media : mediation as a vital process / Sarah Kember and Joanna Zylinska.
p. cm.
Includes bibliographical references and index.
ISBN 978-0-262-01819-7 (hbk. : alk. paper)—978-0-262-52746-0 (pbk.) 1. Mass media and technology—Social aspects. 2. Digital media—Social aspects. 3. Social media. I. Zylinska, Joanna, 1971–. II. Title.
P96.T42K46 2012
302.23—dc23
2012008439

10 9 8 7 6 5 4 3 2

Contents

Media, Mars, and Metamorphosis (An Excerpt)[1]

The author Jeremy Hoyle is a former student, and at times zealous disciple, of Francis Fukuyama. His work echoes and extends the concerns Fukuyama expressed in Our Posthuman Future *for the status of human nature in the era of biotechnology and for the rights of the individual in a threatened liberal democracy. Like Fukuyama, Hoyle considers himself a social philosopher, and he too is something of a populist. For his forthcoming book,* Media, Mars, and Metamorphosis, *he has sought out three of the most recent and controversial experiments in biotechnology in order to dramatize his concerns; each promises (or threatens) to change the meaning of human life. Hoyle's chosen experiments incorporate bacteriology, immunology, and what—in the service of rhyme rather than reason—we might call mediology. The experiments occupy different spatial realms that Hoyle considers to be analogous: cosmic space, the interior space of the computer, and bodily space at the boundary between self and other. Each experiment—and this is what makes Hoyle's book remarkable—has already been deemed successful, so the following inventions, discoveries, and innovations are therefore highlighted: (1) an experiment designed to test for the presence of microbial life on Mars; (2) an experiment designed to induce tolerance, and therefore eliminate the need for immunosuppressant drugs, in facial transplant surgery; (3) a user-based experiment designed to test the efficacy of, and future prospects for, intelligent media.*

Hoyle, like Fukuyama, is drawn to headline-grabbing events and opportunities. He wants to be a spokesman for ordinary people who are interested in the changing world around them and who have legitimate concerns about the extent to which those changes are good or bad. Although he recognizes the importance of progress in scientific and technological research, Hoyle is concerned that these experiments have gone too far and crossed the line protecting the sanctity of human identity; that, told from a personal perspective, they may not have been as successful as they initially appeared; and that the experiments have not necessarily produced anything new. The Martian microbe is essentially the same as its Earth-based counterpart; the human body always rejects invasion; and research into intelligent media has learned the lesson from failed research into artificial intelligence and is now overtly human centered. In other words, these experiments were dangerous but ultimately self-defeating. With the transgressive potential of science thus contained, the ubiquity of liberal humanism and democracy is assured, and Hoyle has the questionable privilege of rescuing Fukuyama's retracted declaration of the end of history through his realization that nothing in fact changes.

What is more, as he progresses through each account, Hoyle becomes increasingly skeptical about its authenticity. Where are the follow-up experiments and observations on the release of the green bacteria/microbe? Why the lack of public response? And why are there no images of the face transplant patient in the media? Finally, the whole idea of intelligent media is surely just an oxymoron. Hoyle's narration reaches this moralist and expedient but not illogical conclusion when events in his own life—specifically, his health—take an unexpected turn. He is forced to add, in an endnote, that he has been afflicted by a terrible stomach bug, the relevant detail of which is that its issue—to the bemusement and concern of his doctors—is

"If It Reads, It Bleeds" (2010).

green. He is also convinced that in the course of writing this book, his face has changed almost beyond recognition. At first he tried to put it down to stress, weight gain, sudden aging (we all know that writing can take its toll). But he doesn't look stressed, fatter, or older. He looks different. Worse still, and this has to be a delusion, a sign of sudden mental as well as physical deterioration, is that the usually inert objects that populate his home have started talking to him—the toilet, the mirror—and there seems to be no way of stopping them . . .

1. This epigraph to the book—which signposts a number of key issues that *Life after New Media* engages with—is a customized excerpt from Sarah Kember's short story "Media, Mars, and Metamorphosis" (originally published in *Culture Machine* 11 [2010]) and a still from Joanna Zylinska's video project "If It Reads, It Bleeds" (2010).

Acknowledgments

We would like to thank our colleagues at the Goldsmiths Media and Communications Department and our students on the master's program in Digital Media, which we have been running for nearly a decade, and on the After New Media course, for offering us many provocations and challenges with regard to *how to think* about "new media"—and *what to do* with them. We are also grateful to many other universities that have lent hospitality to our ideas: CRASSH at Cambridge, Johns Hopkins Center for Advanced Media Studies, Media@McGill, Sussex University, Department of Film and Television Studies at Warwick, the Visual Culture program at Westminster, as well as to the Advanced Cultural Studies Institute of Sweden, the Photographers' Gallery in London (in association with the Institute of Modern and Contemporary Culture at the University of Westminster), and the Society for Existential Analysis at the Wellcome Collection. We owe a big personal thank you to Darin Barney, Gary Hall, Janis Jefferies, Joanne Morra, Nina Sellars, Marq Smith, Stelarc, Liz Vasiliou, and Bernadette Wegenstein. We are very grateful to the anonymous MIT Press reviewers, and to Doug Sery and Katie Helke Dokshina at the MIT Press for all their help and support with this project. Last but not least, thanks to everyone at *Culture Machine* for still flying the experimentation flag.

Introduction: New Media, Old Hat

In *Life after New Media*, we set out to examine the current debates on "new" or "digital" media. In doing so, we want to make a case for a significant shift in the way new media is perceived and understood: from thinking about "new media" as a set of discrete objects (the computer, the cell phone, the iPod, the e-book reader) to understanding media predominantly in terms of processes of mediation.

The argument developed in our book, as reflected by its title, *Life after New Media: Mediation as a Vital Process*, is threefold:

1. In an era when being on Facebook or Twitter, having a smartphone or a digital camera, and obtaining one's genetic profile on a CD after being tested for a variety of genetic diseases has become part of many people's lives, we maintain that there is a need to move beyond the initial fascination with, and fear of, "new" media—and beyond the belief in their alleged "newness," too.
2. There is also a need to look at the interlocking of technical and biological processes of mediation. Doing so quickly reveals that life itself under certain circumstances becomes articulated as a medium that is subject to the same mechanisms of reproduction, transformation, flattening, and patenting that other media forms (CDs, video cassettes, chemically printed photographs, and so on) underwent previously.[1]
3. If life itself is to be perceived as, or, more accurately, *reduced to* a medium, we need to critically examine the complex and dynamic processes of mediation that are in operation at the biological, social, and political levels in the world, while also remaining aware of the limitations of the stand-alone human "we" that can provide such a rational critique.

Yet is this proposed move "beyond new media" not a little premature? It was barely a decade or so ago that a new disciplinary alignment emerged at the crossroads of the arts, humanities, and social sciences that was given the name "new media studies"—although the use of the term "new media" can be traced much further back, at least to Harold Innis and Marshall McLuhan.[2] The first phase of "new media studies" was predominantly focused on technology's function in new media platforms and devices (the use of the Internet by children, the global spread of mobile telephony, etc.) and on a radical division between analog and digital media (letters vs. email, film vs. CCD camera sensors[3]). Understandably, much energy during that first phase was spent on developing descriptions and definitions—concerning what these new media really did, how new they actually were, and how they differed from "traditional" or "broadcast" media. It should be noted that the question of the relation between media and technology was elided in many of those debates, a state of events that resulted in the frequent conflation of "new media" and "new technology." Media also tended to become equated with the computer—or, to cite Lev Manovich, "media became new media"[4]—thus erasing the specificities of, and distinctions between, existing old and new media. Entities such as data and information, and processes such as interactivity, convergence, and digitization, became the focus of the rapidly developing discipline of "new media studies."

Many theorists of new media have attempted to make a mark in this emerging field by setting themselves against its earlier definitions and proposing ways to move on and beyond them. For example, Wendy Hui Kyong Chun, one of the editors of the anthology *New Media, Old Media*, argues against a noncritical adoption of the "new media" term by saying, "The moment one accepts new media, one is firmly located within a technological progressivism that thrives on obsolescence and that prevents active thinking about technology-knowledge-power."[5] Yet Chun does not recommend abandoning the term altogether. Instead, she recognizes that "new media" has already been consolidated into a field with its own emerging canon and institutional space. At the same time, Chun argues strongly against perpetuating the myth of the singular uniqueness of new media, insisting instead that the new "contains within itself repetition."[6] To a certain extent, it can be argued that "new media" was already born as a problem, and that the majority of the theorists who have used this term have always done so somewhat reluctantly, with a sense of intellectual compromise they are having to make if they want their contribution to be recognized as part of a particular debate around technology, media and newness. Through running the master's program in Digital Media at Goldsmiths, University of London, and through working on our own publications in the field of "new media studies,"[7] we have become increasingly aware of both the disciplinary seductions and the conceptual limitations of this term.

Generally speaking, scholarship in media studies fits into two methodological frameworks. Those from the social sciences and communications-based disciplines typically approach the media through a mixture of empirical research and social theory, with questions of political structures, economic influences, social effects, and individual agencies dominating the debate. Those from the humanities in turn predominantly focus on what different media "mean"; that is, they tend to look at media as texts and at their cultural contexts. Of course, there are also those who have never felt comfortable to be pigeonholed in this way and for whom questions of language and materiality, of culture and politics, have always needed to be studied together. (Work undertaken from the perspective of the actor-network theory influenced by Bruno Latour, of the materialist philosophy of Gilles Deleuze, and of science and technology studies has contributed toward blurring the distinctions between the two frameworks, or "camps.")

It is at this point that we enter the debate on new media in our book. However, our aim in *Life after New Media* is to do something other than merely provide an extension or corrective to the current field of "new media studies." Instead of developing an alternative definition or understanding of new media, we propose to refocus the new media debate on a set of processes that have so far escaped close analysis by media studies scholars. In other words, with this book we are not so much interested in moving the debate on new media *on*, but rather in moving on *from* the debate on new media and, in doing so, focusing on the concept of mediation. The distinction is of course primarily heuristic—that is, provisional and strategic—and the purpose of separating mediation from media will be to clarify the relation between them. Mediation does not serve as a translational or transparent layer or intermediary between independently existing entities (say, between the producer and consumer of a film or TV program). It is a complex and hybrid process that is simultaneously economic, social, cultural, psychological, and technical. Mediation, we suggest, is all-encompassing and indivisible. This is why "we" have never been separate from mediation. Yet our relationality and our entanglement with nonhuman entities continues to intensify with the ever more corporeal, ever more intimate dispersal of media and technologies into our biological and social lives. Broadly put, what we are therefore developing in *Life after New Media* is not just a theory of "mediation" but also a "theory of life," whereby *mediation becomes a key trope for understanding and articulating our being in, and becoming with, the technological world, our emergence and ways of intra-acting with it, as well as the acts and processes of temporarily stabilizing the world into media, agents, relations, and networks.*

Our theoretical inspiration for this argument predominantly comes from the work of two philosophers: Henri Bergson (and the materialist-vitalist philosophy subsequently developed by Deleuze) and Jacques Derrida (and his deconstructive thinking around concepts, processes and the ethicopolitical nexus). It is with Bergson and

Derrida that we start approaching media as a series of processes of mediation. This entry point will take us toward the examination of the temporal aspects of media—its liveness (or rather, lifeness),[8] transience, duration, and frequently predicted death. Our primary reason for turning to Bergson is that he allows us to raise questions about the more traditional perception of media as a series of spatialized objects (the iPod, the computer) and also about mediation—that is, multiple, entangled processes of becoming. However, we have to bear in mind that the process of mediation is also a process of *differentiation*; it is a historically and culturally significant process of the temporal stabilization of mediation into discrete objects and formations. In the encounter with Bergson's notion of "creative evolution," Derrida's notion of "différance" functions as a kind of interruption or "cut" to the incessant flow of mediation, facilitating as it does the discussion of the symbolic and cultural significance of this interruption. The negotiation between the Bergsonian (or perhaps, more appropriately, Bergsonian-Deleuzian) and the Derridean philosophical traditions is nevertheless of interest to us here only as far as it allows us to think, move with, and respond to the multiple flows of mediation. It is not therefore an intellectual exercise in its own right, just as the book is not *about* Bergson, Deleuze, or Derrida in any straightforward way. Our attempt to read media as "mediation," both critically and creatively, is informed by a rigorous playfulness toward philosophy, borrowed from the long line of feminist critical thinkers such as Donna Haraway, Rosi Braidotti, and Karen Barad, or, indeed, from Bergson, Deleuze, and Derrida themselves. As well as drawing, specifically, on Bergson's intuitive method, we recognize our allegiance to what Braidotti terms a "nomadic, rhizomatic logic of zigzagging interconnections."[9] The latter logic manifests respectful irreverence toward one's predecessors. Resisting the injunction to speak in our masters' or mistresses' voices, we are therefore seeking methods of thinking and writing that can allow us to see and make a difference.

One of the central issues that concern us in this study of the temporal aspects of media is the relation between events and their mediation. Our argument is that events are never merely presented and *re*presented in the media, and that any such representations are always to an extent performative. Philosophers such as Derrida and Bernard Stiegler, as well as many media scholars, associate media—especially television—with the *illusion* of liveness. Liveness is particularly linked with television news and the coverage of disaster and catastrophe. Generally, it is regarded as a sleight of hand. Yet if we regard such illusory liveness as performative—that is, as being able, to an extent, to bring about the things of which it speaks; things such as "the credit crunch" or "war on terror," say—then not only will we be able to explore questions such as "Did Robert Peston (BBC Business Editor) cause the recession in the UK?" but we will also avoid a reading of media that is overly constructionist, static, and—ultimately—lifeless.

As a continuation of the previous argument, we suggest that mediation gives us insight into *the vitality of media*. By the latter, we mean something more than just the *liveness* of media that we know about through television studies of catastrophes and other "newsworthy" occurrences. We are referring instead to the *lifeness* of media—that is, the possibility of the emergence of forms always new, or its potentiality to generate unprecedented connections and unexpected events. This issue raises the following set of questions for us: if we are saying that the events we have looked at are, to differing extents and in different ways, performed through their mediation, then how should we respond to them in our critiques? Are our critiques not also forms of invention? Or, more broadly, can we think of a way of "doing media studies" that is not just a form of "media analysis" and that is simultaneously critical *and* creative? Could it allow us to challenge the opposition between "media theory" and "media practice" that many university media departments have adopted somewhat too comfortably over the years, at worst privileging one over the other, at best aiming at some kind of dialectical resolution that in the end only reaffirms the division?

In the light of such an argument, any attempt to root media analysis in fixed entities such as "the social," "subjectivity," "economy," "politics," or "art" must therefore be seen as nothing more than a pretense. It is not that many traditional forms of media analysis do not recognize the need for this pretense. Nevertheless, what *Life after New Media* argues is lacking in many such analyses is a serious engagement with the consequences of this recognition, in ways that would be both critically rigorous and adventurously inventive. This is perhaps an appropriate moment to insert a personal double confession into our introduction. The writing of this book has coincided for us with the consolidation of our longstanding ambition to enact knowledge production and media production differently: Sarah Kember has a literary agent and has published her first novel, *The Optical Effects of Lightning*, and Joanna Zylinska has completed a master's degree in fine art photography and started exhibiting her work. Yet at the same time, these incursions into what academic conventions traditionally designate as "practice" have reaffirmed our commitment to rigorous scholarship and to attentive readings of texts and concepts—even if they have pushed further our desire for experimentation and boundary crossing. By drawing on different instances of media enactment, we thus hope to have outlined in this book a more dynamic, networked, and engaged mode of working on, in, and with "the media," in which critique is always already explicitly accompanied by the work of participation and invention. *Life after New Media* closes off with our proposal for "creative mediation" understood as a mode of "doing media studies" otherwise. The book thus emerges out of a complex system of intertwined intellectual, social, economic, and artistic influences that have been shaping the interdisciplinary field of new media studies for nearly two decades now and that have been shaping us as scholars, writers, and teachers within this field. It is an experiment in producing knowledge differently, in exercising academic borrowing and hospitality, in asking questions about "media production" of both ourselves and others, in literally writing and thinking in multiple voices and tongues. As well as providing a name for the ever changing mediascape, mediation for us stands for this dynamic entanglement of ideas, voices, and minds.

Chapter 1 makes a case for a shift from thinking about "new media" as a set of discrete objects to understanding media, old and new, in terms of the interlocked and dynamic processes of mediation. It also outlines what is at stake in this shift from thinking about media solely as objects of use, to recognizing our entanglement with media not just on a sociocultural but also on a biological level. Introducing the work of the philosophers Martin Heidegger and Bernard Stiegler, we read mediation as an intrinsic condition of being-in and becoming-with the technological world. We then offer to see mediation as the underlying, and underaddressed, problem of the media.

If "the media narrate ordinary life by anticipating it, with such force that its story of life seems ineluctably to precede life itself," for the philosopher Bernard Stiegler, public life is actually "produced by these [media] programs."[10] Chapter 2 focuses on two media "events," or "crunches," that are linked by the prospect of global or even cosmic disaster: the "credit crunch" of 2007–2009 and the "big crunch," otherwise known as the Large Hadron Collider (LHC) Project at CERN, Switzerland. Although the latter was purposefully designed in 2008 with a view to recreating the conditions that prevailed immediately after the Big Bang, public apprehension has centered on the possibility that black holes will be formed, signaling the end of the world. As the experiment in particle physics that stresses the contiguous nature of space-time at the origin of life, the universe and everything else, the LHC project offers perhaps the *definitive event* by means of which we might effectively intuit the process of mediation—or the existence of life after new media.

Since the event of mediation is, like time (or, indeed, life itself), both invisible and indivisible, any attempt at its representation must ultimately fail. In chapter 3, we offer a challenge to representationalism by looking at photography, its historical ambitions, and its various techniques. Photography is understood here as a process of cutting through the flow of mediation on a number of levels: perceptive, material, technical, and conceptual. The recurrent moment of the cut—one we are familiar with not just via photography but also via film making, sculpture, writing, or, indeed, any other technical practice that involves transforming matter—is posited here as both a technique (an ontological entity encapsulating something that *is*, or something that *is taking place*) and an ethical imperative (the command: "Cut!"). The key question that organizes our argument is therefore as follows: if we must inevitably cut, and if the cut functions as an intrinsic component of any creative, artistic, and especially photographic practice—although this is still only a hypothesis—then what does it mean to *cut well*? In introducing a distinction between photography as a practice of the cut and photographs as products of this process of cutting, we also aim to capture and convey the vitality of photographic movements and acts.

Chapter 4 compares media visions of the transnational or even cosmic future discussed in the earlier chapters with the viewing point of the domestic present. Arguably, our homes, like our bodies, have always functioned as "intelligent" media. They foreground location and identity as a counterforce to dislocation and differentiation. This set of associations is clearly reflected in the idea of "the smart home," which is embedded with networked computational objects and speech-based autonomous agents who travel so that we can remain in place, safe and protected from a hostile environment. The smart home promises mobility without movement, and fulfills "a long-standing dream of artifacts that know us, accompany us"[11] and comfort us. Intelligent mediation, centered increasingly on the home, is not, as it is sometimes presented, about celebrating hybrid human-machine agency. It is more about positioning "us" as threatened but ultimately reassured subjects, with our private, individualized patterns of media consumption. We argue that intelligent mediation thus becomes a facet of neoliberalism, functioning as the reinforcement of self-interest in the face of both alterity (of what, in a cosmic sense, we might become) and adversity (or what, in the more immediate economically prescribed future, might become of us).

In our attempt to envisage different sociopolitical contexts and different futures, in chapter 5 we explore the possibilities of a less conservative, more inventive approach to the mediated self. It is premised upon a rupture with neoliberal logic and with the reaffirmation of a unitary, autonomous, and authentic subject—a rupture enacted by taking the issue of time and its passage more seriously. The prospect of self-mediation also redefines stability in terms of the inevitable limitations of becoming. In this chapter, we will explore the limitations of transformative self-mediation through a reading of cosmetic surgery (including extreme surgical transformations and the normalizing role of makeover TV shows) and face transplant surgery. Our reading is consistent with a posthumanist, particle physics–based approach informed by theorists such as Karen Barad. If facial surgery is an instance of biotechnological self-mediation writ large, because it is literally inscribed on the body as a medium, then self-mediation is a process that moves "us" both home and away, consolidating and authenticating our experience even as it extends and imperils our identity.

Chapter 6 pursues the ethical implications of this ultimate instability and transience of the mediated cultural subject. It investigates what exactly is entailed in the recognition that "nobody and no particle of matter is independent and self-propelled, in nature as in the social."[12] It also asks what moral frameworks become available within the context of ongoing dynamic mediation, and whom ethical responsibility concerns if we are all supposedly "becoming Facebook" (no matter whether we are "on" it or not). In the light of the above, we outline what we term "an ethics of mediation"—which, in line with our expanded understanding of mediation as a way of being and becoming in the technological world, with all its biodigital configurations—can also be dubbed "an ethics of life."

Positioned as a kind of critical summary, chapter 7 engages with the idea of "creativity" in the context of both life's supposed creative potential and the work on creativity from the context of creative industries, in preparation for our attempt to offer a different mode of doing critical work "after new media." Such a mode is indicated in Bergson's intuitive method and is echoed in the work of many feminist philosophers. The second part of this chapter adopts the format of a "live essay" in which one of the crucial oppositions in media studies—that between "theory" and "practice"—becomes a subject not only of critical interrogation but also of a performative event. Drawing on our own media practices (creative writing and photography, respectively), we hope in this way to have taken some steps toward enacting, rather than just proposing, "life after new media."

1 Mediation and the Vitality of Media

False Problems and False Divisions

This chapter makes a case for a shift from thinking about "new media" as a set of discrete objects to understanding media, old and new, in terms of the interlocked and dynamic processes of mediation. It also outlines what is at stake in this shift from thinking about media solely as things at our disposal to recognizing our entanglement with media on a sociocultural as well as biological level. This argument will lead us to pose the following question: if media cannot be fully externalized from subjects, or "users," then how might "we" engage with "them" differently? We will also consider the political and ethical implications of such engagements.

After outlining the key debates on new media within media, communications, and cultural studies, we will turn to the work of philosophers Martin Heidegger and Bernard Stiegler to explore the relationship between "media" and "technology" and to advance a proposition that mediation is an intrinsic condition of being-in, and becoming-with, the technological world. With this proposition, we will offer to see mediation as the underlying, and underaddressed, problem of the media. As the role of this chapter is first of all to provide a theoretical framework—a toolbox of concepts we will be working with throughout the course of this book—we will also seek to distinguish between the question of mediation and the question of media. This distinction is primarily heuristic—that is, tentative and pragmatic—and the purpose of separating mediation from media will be to clarify the relation between them. Henri Bergson's philosophical method of division and reintegration, as reappropriated by Gilles Deleuze, will be of particular use to us here. This "method" proposes three things: (1) that we distinguish between "true" and "false" problems, (2) that we distinguish between differences in degree and differences in kind, and (3) that we consider the object of our inquiry in terms of its temporality.[1] This last law, or rule, is the most important one for Bergson, and it will be the principal means by which we will seek to distinguish between media and mediation.

Having offered a preliminary investigation of the concept of mediation, we will then present mediation as the underlying and underaddressed problem of the media. We will do so by highlighting, and then bracketing, the "false" problems and false divisions associated with debates on new media. To continue with our use of the Bergsonian heuristic, these problems and divisions are "false" not in any ontological sense related to some originary idea of truth, but rather because they limit the understanding of the complex and multifaceted phenomena and processes by imposing clear-cut distinctions and categories all too early. This process of fragmenting the world into particular categories, often arranged into sets of oppositions, is not only reductive and therefore unhelpful; it also has serious political and ethical consequences for our understanding of the world, its dynamics, and its power relations. Thinking through and against such false problems and oppositions is therefore also a political intervention into "the media"—one that is different from studies of the political economy of media and communications, for example, but that is not any less serious or important.[2] In addition to the false problems (which we identify in discussions on new media that focus on a singular problem, such as newness, digitization, interactivity, convergence, or data, at the expense of all the others), the field of new media is arguably also marred by a number of "false divisions"—or what cultural theorists trained in poststructuralist thought tend to refer to as "binary oppositions." Such false divisions that have so far shaped debates in new media studies include determinism and constructionism; technology and use; theory and practice; structure and agency; information and materiality (an extension of the division between language and materiality); and subjectivity and objectivity.

Even where these false divisions have been identified as such—and of course many writers are aware of their limited currency—it has proven difficult to avoid them.[3] The reason for this difficulty partly lies with the residual effects of disciplinarity and the associated requirement to take a set of key concepts within a given discipline and then elevate them to a transcendental position, as a result of which everything else gets questioned or even dismantled except for these foundational concepts (for example, "data" and "information" in computer science; "subjectivity" in psychology; "society" in sociology). Another reason for the survival of such false divisions lies perhaps in the prevalence of social science perspectives in media, communications, and cultural studies, perspectives that are fundamentally positivist and humanist, and that stake empirical claims on partial perspectives of "black-boxed"—that is, isolated, protected, and simultaneously obscured—aspects of the media. "Politics" and "the social" are just two examples of such privileged terms within the dominant, social sciences–informed tradition of media and communications.

Our own argument in the book is that although *media constitute differences in degree*[4] that should not be elided under any overarching concept, *mediation* nevertheless *constitutes a difference in kind*. It cannot be isolated and hence stabilized in any straightforward manner because its mode is fundamentally that of time. The interdisciplinary nexus of media, communications, and cultural studies—within which questions of new media are most readily addressed—has not so far offered an adequate account of mediation as a process because it has not taken the temporality of media seriously. We aim to address this rather substantial shortcoming in the pages that follow. This chapter will end with a proposition to see mediation as the expression of media temporality, or what we will term the "lifeness" of media.

"What Is New about New Media?" and a Few Other Old Debates

Many commentaries on emerging media—"Everyone who is anyone is on Facebook!"; "Apple has revealed an iPad!"; "the Internet causes obesity in children!"—tend to fall into one of the two extremes: technophilia or technophobia, utopianism or dystopianism, that is, either a celebration of or cynicism about the advent of the supposedly new. Similar sentiments, albeit articulated in a more restrained manner, tend to inform a high number of academic arguments about new media and their supposed influence. This limited dualism, or simple binary or oppositional thinking, is not, however, restricted to *feelings about* new media: it also structures many *ontological conceptualizations of* them (analog vs. digital, closed vs. open, centralized vs. distributed, readerly vs. writerly, mass vs. participatory). The majority of debates on new media thus tend to perpetuate the "false divisions" discussed previously. The old versus new division plays a special role among those oppositions in that it not only brings together affect and matter but also inscribes media into a progressive developmental narrative. In other words, it introduces the question of time into debates on media while simultaneously freezing this question by immediately dividing "media time" into a series of discrete spatialized objects, or products that succeed one another. Thus we are said to progress *from* photography *to* Flickr, *from* books *to* e-readers.

As mentioned in the introduction, the alleged "newness" of the products and processes that get described as "new media" should not be taken at face value—not only because of the rather problematic historical trajectory of progressive media development this narrative adopts, as persuasively argued by Lisa Gitelman in *Always Already New: Media, History, and the Data of Culture*,[5] but also due to the ideological implications of any such designation. We could perhaps go so far as to say that in descriptions of this kind, "newness" functions predominantly as a commercial imperative: it demands that we keep upgrading our computers, cell phones, and communication and data storage devices in order to avoid obsolescence—the obsolescence of both our equipment and of ourselves in a world whose labor and social relations are being presented as increasingly fluid. Defining "new media" in terms of their convergence and interactivity—two key characteristics frequently evoked in relation to this concept—links this imperative for connectivity with the neoliberal fantasy of ultimate consumer choice. The promoted interactivity of devices such as Apple's iPad or Amazon's Kindle is therefore perhaps first of all an expression of the ideology of "added value," one that sets the supposedly passive consumption of "old media" against a more active and engaged consumption of new media. (This argument works better with the iPad as a multimedia device or the computer as an omniproductive machine than it does with Kindle.) We may recall at this point the always already interactive and converged nature of "old media," such as radio, newspapers, or books. Due to their relative longevity, books are a particularly interesting example of "old media." The universe constituted through and within them, be it a philosophical plane of immanence or a fictional world of a novel, has always required an active participation and contribution from the reader, not to mention the efforts of all those who have been involved in their editing, design, production, and distribution. Arguably, books are thus as hypertextual, immersive, and interactive as any computerized media.

As Gary Hall explains in *Digitize This Book!* by drawing on the work of the historian Adrian Johns, until the mid-eighteenth century, the book constituted an unstable object, with Shakespeare's folio, for example, including more than six hundred typefaces and many inconsistencies with regard to spelling, punctuation, and page configuration. "Early in the history of the printed book, then," writes Hall,

> readers were involved in forming judgments around questions of authority and legitimacy: concerning what a book is and what it means to be an author, a reader, a publisher, and a distributor. The development and spread of the concept of the author, along with mass printing techniques, uniform multiple-copy editions, copyright, established publishing houses, editors, and so forth meant that many of these ideas subsequently began to appear "fixed." Consequently, readers were no longer "asked" to make decisions over questions of authority and legitimacy.[6]

However, the inherent instability of the "old medium" of the book never disappeared altogether, according to Hall; it just became obfuscated—and is particularly difficult to see amid the debates about the (inter)active and allegedly more creative and collaborative nature of "new media."[7] Yet the delegation of decision-making processes to computer algorithms (the logic and hierarchies behind Google's search engine; Amazon's recommendations such as "Customers who bought this item also bought X," etc.) raises some serious questions for this professed interactivity of new media. Arguably, it offers a freedom of consumer choice *in place of* any actual interactivity, creative engagement, and what Hall refers to in his book as responsible and ethical decision making.

Significantly, when *The Guardian* reported in July 2010 that according to the online bookseller Amazon, "sales of digital books have outstripped US sales of hardbacks on its Web site for the first time," it also commented that this announcement "will provoke horror among those who can think of nothing better than spending an afternoon rummaging around a musty old bookshop."[8] Even though the article also included the information that the "hardback sales are still growing in the US, up to 22% this year," it was decisively framed through the familiar old-new dichotomy, with the former bringing up associations of mustiness, outdatedness, and not moving on "with the times." While the melancholy if sympathetic figure of the leisurely reader was depicted here as "rummaging," no doubt languorously and somewhat clumsily, through the bookshop, we were supposed to imagine the reader of e-books as whizzing through the pages on her digital device. The latter reader thus won the contest of time, of reading as (inevitable) speed-reading, but also of being au courant with a media practice that requires a new set of technological devices to "keep up" with the times and also requires one to not shy away from what the future "brings."

Questions about future media and about the future of the media ("Will our homes become more intelligent?"; "Will the Internet kill broadcast journalism?"; "Will users become more inclined to pay for online content?") are part and parcel of the debates about the "newness" of media. What is most problematic about futurism in as far as it predicts and speculates about the social, psychological, and economic *effects* of new media is not so much the extent to which those predictions and speculations are accurate. Incidentally, they usually are not. As the Polish writer Stanislaw Lem ironically commented in his *Summa Technologiae*, "Anyone can have some fun by just putting in a drawer for a few years what is currently being described as a believable image of tomorrow."[9] And thus, when Alan Turing predicted in 1950 that by the year 2000 the idea that machines could think would become commonplace,[10] or when Fred Ritchin announced in the early 1980s that digitization would bring about "the end of photography as we have known it,"[11] they were evidently both wrong, but that is not the major problem with those predictions. What is even more problematic from our point of view is that these speculations about media futures were relying on a kind of linear, cause-and-effect thinking that carries the name of *technological determinism*. Marshall McLuhan's prognosis that high-resolution TV would no longer be television, "which depends on seeing as causes the details of a technology that may well change,"[12] fell into a similar trap.

In *New Media: A Critical Introduction*, Martin Lister and colleagues oppose *technological determinism* to *social constructionism* and, in doing so, reveal another limited dualism which underpins new media debates. At the heart of the opposition between determinism and constructionism lies the question of technology and, with it, of technological agency. To briefly recap, technological determinism proposes that technology *causes* changes in culture and society. From this perspective, the Internet is presented as having revolutionized the way we communicate, or, in the words of new media writer Nicholas Carr, "the way we think, read and remember."[13] For others, such as Clay Shirky, media consultant and author of *Here Comes Everybody: The Power of Organizing Without Organizations* (2008) and *Cognitive Surplus: Creativity and Generosity in a Connected Age* (2010), the Internet is said to have transformed the very fabric of society by changing the way people collaborate. Social constructionism (also referred to as constructivism), on the other hand, regards technology as merely the *effect* of ongoing changes in culture and society. Here, the Internet would be positioned as having emerged in order to meet the needs of the American military, with the original ARPANET developed for the purposes of the US Department of Defense as a decentralized military research network aimed at surviving a nuclear strike. In other words, the Internet is seen as only the means to specific governmental and military ends.

Following Lister and colleagues, the question of technological determinism can be traced back to the debate between Marshall McLuhan and Raymond Williams in the 1960s and 1970s. We are making a quick stopover by this well-rehearsed debate because the work of those two thinkers was foundational to and still informs the way media and technology are being talked about and understood in many academic and mainstream debates today. Lister and coauthors demonstrate how the first position gives sole power, or agency, to technology, while the second position awards power and agency to human cultures and societies. One is therefore technicist, the other humanist—but both are deterministic. Determinism, then, can be understood as an attempt to decide "how far new media and communications technologies, indeed, technologies in general . . . actually determine the cultures that they exist within" and conversely "how cultural factors shape our use and experience of technological power."[14]

Lister and colleagues dramatize the question of technological determinism through the work of McLuhan and Williams. If McLuhan is presented as übertechnicist in many media, communications, and cultural studies accounts, then Williams occupies the very antithesis of that position. "While McLuhan was wholly concerned with identifying the major cultural effects that he saw new technological forms . . . bringing about, Williams sought to show that there is nothing in a particular technology which guarantees the cultural or social outcomes it will have," write Lister and coauthors.[15] The question of determinism, then, is set out from the beginning in terms of *all or nothing*. From that perspective, technology is seen as either having major social effects or none at all! The authors of *New Media: A Critical Introduction* inevitably simplify McLuhan and Williams's positions in their attempt to frame technology as a key question for our times. However, in taking recourse to such an inevitably reductionist theoretical model, they draw out a very important point. In the battle between McLuhan and Williams—which is a battle between determinism and constructionism, between technicism and humanism, between machine agency and human agency—it was Williams who won. That is, according to Lister and colleagues, he gained prominence in the context of *British* cultural and media studies—if not within the technicist cyberculture studies, European media theory, or North American mass communications, where the influences may have remained more McLuhanite. (We can also mention here the significance of works by Neil Postman, Friedrich Kittler, or Jean Baudrillard for those latter contexts.) The effect of this somewhat Pyrrhic victory was the loss of the question of technology from within the main trajectories of media, communications, and cultural studies. Consequently, the question of technology in the Williams-inspired, partial, but still influential approach became synonymous with technological determinism, and with McLuhan himself, so it disappeared from debates about culture and media. This disappearance, we might say (a point with which Bolter and Grusin would probably agree), was something of a mistake.

It was a mistake because—banal as it sounds, but this is actually an important philosophical point, which we will expand upon further—media cannot be conceived as anything else than hybrids, and technology is part of that hybridity.[16] This mistake may perhaps be rectified by a return to McLuhan, a return that will work through or even attempt to bracket his alleged determinism, but that will retrieve his valuable foregrounding of hypermediacy and the body. (In many ways, McLuhan's work anticipated Bolter and Grusin's "remediation" thesis.) Lister and colleagues refer to McLuhan's "physicalist" emphasis, that is, his sense of technologies as physical prostheses or extensions of the body. This emphasis, they say, is "precisely what humanism in cultural and media studies has been unable to address,"[17] and it potentially opens up, or perhaps rather reopens, the question of the relation between "biological and technological things."[18] The question of technology is then, simultaneously, the question of biology. It is a question that "we" (or some of us at least), according to the authors of *New Media*, have been unable to ask for too long. Even if the way that Lister and coauthors extrapolate from the British intellectual context to speak about the study of media and culture generally or even universally is problematic, their attempt to refocus the debate on new media around core philosophical questions—those of the ontology of media, and their kinship with other objects or, indeed, life forms—deserves further attention.

Remediation and Its Discontents

Following in Lister and colleagues' footsteps, we want to spend some time looking at Bolter and Grusin's theory of *remediation*. By now a staple in new media studies, both celebrated for its brilliance and criticized for its generality and forced ubiquity, this term refers to the way in which the computer "refashions older media" and in which older media "refashion themselves" in the context of the computer. According to Bolter and Grusin, old and new media co-exist, but one type of media does not necessarily "swallow up" the other. If "immediacy," or realism, is a defining characteristic of old media—in the sense that in a Hollywood blockbuster things "look" like they are real, trying to make us forget for two hours that we are watching a movie; and "hypermediacy," or the collage effect of media forms and styles associated, for example, with the aesthetics of websites is a defining characteristic of new media, then they also, currently and historically, coexist. Remediation offers a critique of the teleology of technological convergence because it does not cut off the present from the past, the new from the old. Old media come around again in this framework, as a result of which history is seen not as linear and progressive but rather as nonlinear and cyclical. After Foucault, Bolter, and Grusin opt for the idea of history as genealogy—that is, the tracing of the contingent emergence of ideas and knowledge systems through time—rather than as teleology, a progressive and purposeful linear development of events.

The concept of remediation obliges us to consider interactive TV, for example, in the context of the history of television, or cell phones in the context of the history of the telephone. Tying new media to old media is, Bolter and Grusin say, a "structural condition" of all media. More than that, though: by incorporating a concept of hypermediacy, remediation requires that we consider one medium and its history in relation to other media (and their histories). Bolter and Grusin write, "Digital visual media can best be understood through the ways in which they honor, rival, and revise linear-perspective painting, photography, film, television, and print. No medium today, and certainly no single media event, seems to do its cultural work in isolation from other media, any more than it works in isolation from other social and economic forces."[19]

We can think here of a number of media events that have been hypermediated in this way at the beginning of the twenty-first century: the flu pandemic, the Arab Spring, or the global credit crunch. As we will argue in more detail in the next chapter with reference to the third of these examples, all of these events reveal a complex relationship between the event "itself" and its mediation, foregrounding the significance of technological, social, economic, geographical, and other influences or forces well beyond those controlled by the human. This complexity demands an interdisciplinary nonhumanist theoretical framework that would facilitate such a multifaceted understanding. As Lem figuratively puts it in his *Summa Technologiae*, a more humanist mode of thinking along the lines: "The human spirit, having experienced failures and successes throughout the course of history, has eventually learned to read from the Book of Nature,"[20] used to be dominant in the fields that provide explanations of science and technology, with media typically being positioned as delimitable entities and effects of the existent situation. What Bolter and Grusin's concept does, then, is open up the possibility of a nonhumanist reading of media as dynamic, complex, and interwoven processes beyond the singular control of the human (even if the authors themselves, as we will show later in this chapter, do not fully follow the consequences of this realization).

Veering between understanding mediation as a hybrid process on the one hand, and presenting it in terms of relatively discrete media on the other, Bolter and Grusin do indeed stop short of embracing the full radicalism of their remediation thesis. In saying that "Our culture wants both to multiply its media and to erase all traces of mediation,"[21] they end up reducing culture to a unified being with its own rather uncomplicated volition. This anthropomorphization of culture is only one problem with their argument. Another lies in the ascription of relatively transparent affects and effects to human actions. And thus, when discussing the automation of the technique of linear perspective in media such as photography, film, TV, and computing, they say, "A photograph could be regarded as a perfect Albertian window" to the point of both concealing the process and eliminating the artist, thus offering "its own route to immediacy."[22] Drawing on the work of André Bazin and Stanley Cavell, they argue that all of these media forms are aimed at satisfying our "obsession with realism" and "our culture's desire for immediacy."[23] The media are thus seen here as fulfilling a particular socio-cultural need, and as arriving only in response to this already preformed need. Seemingly conflicted about whether to position desire for media development in individual subjects or in this nebulous entity called "culture," Bolter and Grusin also introduce some degree of hesitation with regard to what and who these emergent media are actually responding to. Yet, in tracing the double logic of remediation that both multiplies media and erases traces of mediation, they ultimately remain too focused on human intentions and desires—as evident, for example, in their declaration that "Today as in the past, designers of hypermediated forms ask us to take pleasure in the act of mediation"[24]—and not enough on effects and acts of the media themselves. In the multilayered processes of remediation, they locate agency firmly on the side of the human, with the media environment receding into the role of a background for a human engagement with media objects. We could say that the history of media is for them a consequence of an ongoing practice of a guild of illusionists, all focused on confusing or deceiving us with their box of tricks: "the goal of the computer graphics specialist is to do as well as, and eventually better than, the painter or even the photographer."[25] In saying this, Bolter and Grusin thus erase any traces of technology as *tekhnē*, that is, art or craft, creation, and hence as artifice, from the production of these images.

Consequently, they end up reducing the very process of mediation to actual media—transparent or hypermediated—which then function as a third party in the subject's desire or quest for authentic experience and, ultimately, as a validation of an autonomous self. As their book in many ways provides an inspiration for our study of mediation as both a key process and underaddressed problem within media, communications, and cultural studies, it is important for us to recognize the inconsistencies in their account. However, it is equally important to highlight the conceptual opening made by their focus on media dynamics that goes beyond singular media objects. Witness the following account they provide: "*All* mediation is remediation. . . . at this extended historical moment, all current media function as remediators. . . . Our culture conceives of each medium or constellation of media as it responds to, redeploys, competes with, and reforms other media."[26] Remediation becomes for Bolter and Grusin a name for the "*mediation of mediation*," with each act of mediation depending on other acts of mediation. This leads them to posit the "*the inseparability of mediation and reality*."[27] Referring to poststructuralism and in particular to the work of Jacques Derrida, the two authors acknowledge that there is nothing *prior to* mediation. Yet their argument takes a rather surprising turn when they bring in a somewhat loose and hence problematic notion of the "real" (in which sometimes it stands for the viewer's experience and at other times for entities existing in the world) to declare that "all mediation remediates the real. Mediation is the remediation of reality because media themselves are real and because the experience of media is the subject of remediation."[28] Needless to say, this conclusion, with its rather fixed notion of what counts as "the real," "subject," and "experience," undoes any poststructuralist intimations of their earlier argument—only then to be redeemed, perhaps in a truly poststructuralist manner, by a rather significant footnote (or, to give the MIT Press designers their due, "sidenote"), in which the authors of *Remediation* claim that "media and reality are inseparable."[29]

Significantly, although it does pay attention to early media forms, the concept of remediation does not pin new technologies back onto their historical antecedents in order to conduct a comparative analysis: instead, it considers both change and continuity of media. The question of *what emerges* through processes of remediation is as important as the question of *what is being remediated*. Thinking about the media in processual terms creates problems for the traditionally more dominant humanist approach discussed earlier, whereby the question of technology was elided in favor of the question of its human use. This humanist approach gave rise to arguments such as those by Carolyn Marvin in her book *When Old Technologies Were New*. Marvin's argument is notable as far as it creates problems for technologically deterministic ways of thinking by tracking new technologies such as the computer back to their origins—in this case, in the Victorian telegraph. She writes, "In a historical sense, the computer is no more than an instantaneous telegraph with a prodigious memory."[30] But Marvin does not tell us enough about the singular significance of that prodigious memory or about the material differences and the specific affordances of the computer and the telegraph. She does provide a thorough account of the social history and the uses of the media that were considered "new" toward the end of the nineteenth century (telegraph, telephone), but says arguably too little about their status *as technologies* and about their limited though not negligible agency. The statement she makes that "The history of media is never more or less than the history of their uses"[31] is representative not only of her approach but also of the broader Williams-inspired intellectual trend in media, communications, and cultural studies discussed previously. This statement assumes a one-way traffic from media as discrete objects to humans as their masters, producers, and users. Indeed, Marvin insists that a historical account needs to lead us away from *media themselves* "to the social practices and conflicts they illuminate."[32]

Gitelman's *Always Already New* deserves a mention here as both a continuation and an overcoming of "the Marvinesque perspective." This is evident in Gitelman's statement that "media are curiously reflexive as the subjects of history"—which is another way of saying that "there is no getting all the way outside or apart from media to 'do' history to them; the critic is also always already being 'done' by the media she studies."[33] Reluctant to grant any essentializing agency to "the media," Gitelman correctly points out that media are nothing without their human "authors, designers, engineers, entrepreneurs, programmers, investors, owners, or audiences."[34] However, in her upfront repudiation of "the idea of an intrinsic technological logic," she is less inclined to study this media-user entanglement from its other side, that is, to explore to what extent and in what way "human users" are actually formed—not just *as users* but also *as humans*—by their media. Although the concept of "media" is helpfully loosened up in Gitelman's argument—when she claims that it is a mistake to write about "the telephone," "the computer," or "the Web" as if they were unchanging objects with self-defining properties—she is still prepared to grant "the media" a relatively stable ontology as long the object has been adequately isolated and historicized (or, as Bergson would have it, "solidified"): "telephones in 1890 in the rural United States, broadcast telephones in Budapest in the 1920s," and so on.[35] For Gitelman, then, (a post factum) "specificity" offers a way out of the instability dilemma.

Many thinkers working at the boundaries of media theory and computing have taken significant steps toward arguing for the need to study the "media themselves" and the dynamic logic they activate, rather than focusing on questions of their *human* representation, *human*-centered meaning, or *human* use. For example, in her study *My Mother Was a Computer*, Katherine Hayles develops her earlier ideas on medium specificity and looks at the relationships between language and code in different media. Acknowledging the significance of Bolter and Grusin's notion of remediation, she nevertheless criticizes them for locating the starting point for the cycles of immediacy and hypermediacy "in a particular locality and medium."[36] Hayles proposes instead the concept of *intermediation*, which promises to examine agency at the boundary between biological and technological things, human users and their computers. Karen Barad's concept of *intra-action*—as opposed to *inter*-action—goes even further in that regard because it recognizes that there is actually no "between" as such and that human and nonhuman organisms and machines emerge only through their mutual co-constitution.[37]

This essentially McLuhanite emphasis on the connectedness rather than isolation of media as developed in different theories of re-, inter-, and intramediation has led some commentators to propose that we are currently living in a "media ecology."[38] Even though we recognize the logic behind this assertion, in our book we focus more on the temporal aspects of media that remediation foregrounds rather than on the more frequently discussed spatial or environmental ones.[39] By highlighting the dynamics of media, the concept of remediation paves the way for an examination of processes of *mediation* that are complex and heterogeneous and that take heed of what McLuhan termed the "all-at-onceness" of the "world of electric information."[40] This "all-at-onceness" poses a challenge to many forms of conventional media analysis—in which problems such as technology, use, organization, and production are frequently studied in isolation.

Although the question of technological use should not be discounted, it cannot be given primacy if, *with* Bolter and Grusin, but perhaps also *against them*, we regard remediation as a process which incorporates multiple agencies (technologies, users, organizations, institutions, investors, and so on). As Lister and colleagues point out, Williams's notion of use may work well enough for the analysis of a given technology, but it works less well "if we consider the extent to which technology becomes environmental."[41] It is not simply the case that "we"—that is, autonomously existing humans—live in a complex technological environment that we can manage, control, and use. Rather, we are—physically and hence ontologically—part of that technological environment, and it makes no more sense to talk of *us* using *it*, than it does of *it* using *us*. McLuhan therefore had a point when he argued that, as Lister and coauthors put it, "the human sensorium is under assault from the very media into which it extended itself."[42]

Originary Technicity, or We Are Media[43]

If we take this process of technological extension seriously enough—not just on the level of theoretical argument but also through our experiential being with technologies and media such as cell phones almost permanently attached to our ears, pacemakers, virtual reality goggles, human growth hormones, or Botox—we are obliged to recognize that we human users of technology are not entirely distinct from our tools. *They* are not a means to *our* ends; instead, they have become part of us, to an extent that the us/them distinction is no longer tenable. As we modify and extend "our" technologies and "our" media, we modify and extend ourselves and our environments. This position requires a better, more philosophical understanding of "technology," beyond that of "a neutral tool for the accomplishment of pre-given ends . . . judged by an economic criterion of efficiency."[44] For this, we need to turn to the work of the German philosopher Martin Heidegger.

In his oft-cited essay "The Question Concerning Technology," Heidegger bemoans the fact that in modern times technology has been reduced to the way for humans to organize (or "enframe") nature and bring it under their command. This way of ordering "demands that nature be orderable as standing-reserve."[45] This process of enframing is also a "concealing" because, for Heidegger, it removes us from the more creative, less instrumental relationship with technology we had in the past (which this particular philosopher, like many others, locates in ancient Greece). The historical accuracy of this designation aside, what Heidegger achieves by this excursion to the Greeks is excavating the original meaning of technology as *tekhnē* and *poièsis*, that is, bringing-forth and presencing.[46] Technology for Heidegger is therefore an inherently world-forming process, both on a biological and cultural level. He writes, "Through bringing-forth the growing things of nature as well as whatever is completed through the crafts and the arts come at any given time to their appearance."[47] Yet this process, and human participation in it, are now seriously constrained. In the words of Mark Poster, "As a result of the unconscious quality of modern humans' relation to their framing of things, they do not perceive the setting up of the scene in which they act and take their own cultural shapes."[48] Consequently, "our own being in the world is invisible to us."[49] It is Heidegger's ambition to restore our original relationship with technology, which will also be a way for us to live a more conscious and more free life.

Significantly, when Heidegger declares that "the essence of technology is by no means anything technological,"[50] he is not retreating to some kind of pretechnicist utopia where man used to be at one with "nature." For him, we always remain "chained to technology, whether we passionately affirm or deny it."[51] This is more than just a comment on our spiritual and physical slavery to the forces beyond us: as taken up by Bernard Stiegler's rereading of Heidegger (which we will discuss shortly), this notion points to our *originary technicity*, our way of being-with and emerging-with technology. It is in seeing beyond the instrumental dimension of technology that the human can establish a better relationship to it or even to see himself as part of the technological set up for the world. The essence of technology for Heidegger lies in what he terms a "revealing" of the potentiality of matter, that is, of whatever "does not yet lie here before us, whatever can look and turn out now one way and now another."[52]

Yet, according to Poster, Heidegger's understanding of technology applies more to older technologies represented by machines such as the hydroelectric plant than to information technologies enacted by, for example, the Internet. (To grasp the latter, Poster turns to Deleuze and Guattari, and in particular to Guattari's *Chaosmosis*.) Even if the information age can indeed be dubbed, to cite Poster, "the new order of humachines," it still does not absolve us—technologically dependent humans—of the responsibility of having to figure out ways of being in this humachinic world, with its hybrid ontologies and uncertain ethics. More interesting, perhaps, in worrying about "modern humanity's way of being,"[53] Heidegger can be said to offer a double opening—beyond the humanism of technological use (whereby the perception of technology as a means to an end is only a "fixing," not an "essence")[54] and beyond the determinism of technological "destining." His declaration that the human is "challenged, ordered" to "exploit the energies of nature"[55] implies that the human belongs "even more originally than nature within the standing-reserve." Heidegger is thus prefiguring here Stiegler's thesis of originary technicity. There is technological force at work in the universe that is a challenging and that "gathers man into ordering."[56] As a result, the human has to "respond . . . to the call of unconcealment even when he contradicts it."[57] This having to respond to the environmental force preceding his own existence not only places the human in the condition of dependency and coemergence: it also turns the human into an ethical subject, where ethics is understood as response to what is differentiated from the self and what simultaneously exerts a demand on the self.[58]

In a similar vein to Heidegger, in his *Technics and Time, 1*, Stiegler goes back to the Greeks—in particular, the myth of Prometheus—in an attempt to retrace the history of technological development and thus retell and reimagine what technology means, and what kind of relation we can have with it. The story of Prometheus is for him not just about playing with dangerous objects—fire, weaponry—or making "dangerous discoveries," but rather about the willingness to challenge the established ontological and epistemological order in which man is positioned as a self-contained being, fully present to himself. The myth supposedly illustrates man's technical being. Technology is positioned here as a force that brings man forth and is fully active in the process of hominization: it is not just an external device that can be picked up, appended, and then discarded at will. For Stiegler, the human drive toward exteriorization, toward tools, fire, and other prostheses—toward *tekhnē*, in other words—is due to a technical tendency that is embedded in the older, zoological dynamic, a tendency that was arguably already identified by Heidegger. It is due to this inherent tendency that the (not-yet) human stands up and reaches for what is not in him. It is also through visual and conceptual reflexivity—seeing himself in the blade of the flint, memorizing the use of the tool—that he emerges as always already related to, and connected with, the environmental matter that is not part of him.

In both *Technics and Time, 1* and an interview that constitutes part of the film *The Ister* Stiegler provides a careful exposition of the Platonic dialog *Protagoras*. The dialog tells the story of the creation of mortal beings, including man, and the role that two Greek gods, Prometheus and Epimetheus, played in this process. Epimetheus—a god that Stiegler presents to us as quite absent-minded and not particularly clever—takes it upon himself to furnish all newly created earthly creatures with "qualities." So he distributes strength to the lion, speed to the gazelle, and hardness to the turtle and its shell, making *"his whole distribution on a principle of compensation, being careful by these devices that no species should be destroyed."*[59] By the time he gets to man, however, Epimetheus discovers he has run out of qualities, leaving man unprovided for—"naked, unshod, unbedded, and unarmed."[60] This is the moment when Prometheus comes to the rescue by offering to steal from Hephaestus and Athena the gift of skill in the arts, coupled with fire—"for without fire there was no means . . . for anyone to possess or use this skill." In other words, Prometheus gives man *tekhnē*, while simultaneously completing the creation of the human as a technological being—a being that has the power to create but that also needs to rely on external elements to fully realize his being. Thanks to this newly gained "art," writes Plato, "men soon discovered articulate speech [*phonen*] and names [*onomata*], and invented houses and clothes and shoes and bedding and got food from the earth."[61]

Through his rereading of the myth of Prometheus and Epimetheus, Stiegler provides an alternative story about technology, nature, and the human. Yet he does more than that: he proposes a new framework for man's self-understanding in the technical world. If we really want to get to grips with the question of technology today, claims Stiegler, we must return to the Greeks because they have already framed it very accurately within their own tragic, religious terms. Stiegler's historical excursion therefore has contemporary resonances—not just because Greek philosophy still informs our current notions but also because in that particular myth the Greeks managed to articulate the dramas, tensions, and anxieties of "human becoming" in a world that was constantly evolving. It is in a dynamic, connected model of the world that Stiegler locates the possibility of developing a less hysterical and more responsible understanding of *tekhnē*. What is, however, significant about the *current* moment—and by "current," Stiegler refers to the modern period inaugurated by the Industrial Revolution of the late-eighteenth/early-nineteenth century—is the speed of technological development. It has increased exponentially over the last two centuries, getting out of sync with the speed of the development of other areas of life: social, cultural, spiritual, legal, and so on. This acceleration of the technological development—which is evident in the emergence of machinic production, railway networks, computation, cybernetics, and, last but not least, globalization—has serious consequences for the philosophical order that has been in place since Plato and the Greeks. It is precisely this order that has allowed for the emergence of the hegemonic consensus in modernity, a consensus that sees technics as having no ontological sense, as only an artifice that must be separated from Being.[62] So, even though we have always been technological, a radical change has taken place over the last century, with the speed of technological transformation and intensity of technical production constantly increasing and getting ahead of the development of other spheres of life.

Picking up a Heideggerian thread, Stiegler's work highlights a deconstructive logic at work in the dynamic relation between technology and the human.[63] There is also something unique about the way in which the story of the human as a technical being is told in his early work, which is why we are focusing on this by now quite well known account here. Simply put, in *Technics and Time, 1,* Stiegler seems much more aware that he is telling us a story. He goes back to a number of established oral and written texts not so much with an intention of informing his readers what the world *is like* (a far more dangerous, and, one might argue, hubristically naïve desire that he nevertheless cannot resist in the further volumes of *Technics and Time*) but rather with a willingness to reflect on and think through some of the stories that others have told about the origin of the human: Greek myths, paleontological theories, earlier philosophical accounts. The same stories—including the key narrative about the fault of Epimetheus—are then reframed via another narratological form, that is, the interview with the philosopher included in the video essay *The Ister*. This very act of conscious reiterative storytelling is significant here. The stories about the origin of the human we are told join a long line of technical prostheses such as flint stones and other "memory devices" that have played an active role in the very process of the constitution of the human. In the pre-Platonic, premetaphysical times that the myth of Prometheus and Epimetheus deals with—a myth that Stiegler retells for us—this tragedy is exacerbated by the fact that there is no possibility of redemption from this condition of openness man exists in, other than through the inevitable finality of death.

Yet an interesting breach is created in this theory of originary technicity as outlined by Stiegler, a theory that may perhaps be described—not necessarily in a derogatory manner—as "softly determinist." Drawing inspiration from the work of Emmanuel Levinas and his notion of ethics as originary openness to, and responsibility for, the alterity of the other, as well as from Heidegger's declaration that the human "ek-sists . . . in the realm of an exhortation or address,"[64] we can perhaps go as far as to suggest that this originary technicity is also an ethical condition.[65] If, as Stiegler has it, "the being of humankind is to be outside itself," the always already technical human is a human that is inevitably, prior to and perhaps even against his "will"—productively engaged with an alterity. Being in the world therefore amounts to being "in difference," which is also—for Levinas, as much as for Stiegler—being "in time": that is, having an awareness and a (partial) memory of what was before and an anticipation of what is to come. The idea of the originary self-sufficient, total man living in the state of nature is exposed here as nothing more than a myth, whereby the state of nature stands "precisely [for] the absence of relation."[66] As such, it marks the impossibility of the human (and also of tool use, art, language, and time), as well as of ethics. Originary technicity can thus be understood as a condition of openness to what is not *part of* the human, of having to depend on alterity—be it in the form of gods, other humans, fire, or utensils—to fully constitute and actualize one's being. But this imperative to get outside of oneself and to be technical, that is, to bring things forth, to create, is perhaps also an ethical injunction to *create well*, even if not a *condition* of ethical behavior.

Introducing Levinas—the philosopher whose work is most readily associated with responsibility for the alterity of the human other that makes a demand on the self and that exerts a response from him or her—into our discussion of mediation and technological becoming may seem out of place here. Yet Stiegler's conceptualization of the human as always already technological, and therefore as responding to an expanded set of obligations, offers an opportunity to make Levinas's ethical theory applicable to what we can tentatively call "posthuman agencies."[67] As well as staging an encounter (though not a seamless one) between the Bergson/Deleuze and Levinas/Derrida positions on difference, *Life after New Media* therefore offers an attempt to think about the ethics of connectivity that does not posit any absolute difference of the parties and agents involved in an ethical encounter *in advance* but also recognizes the ethical significance of the process of *cutting through* the seamlessness of life and of the temporal stabilization of agents in this process. Contra Levinas, the processes of material differentiation do not have to be prior to ethical connectivity: the two can rather be seen as dynamically co-constituted. (Chapter 6 considers the ethical questions opened up by the position of originary technicity in more detail; Chapter 7 discusses the significance of creativity as an ethical imperative to "cut well" through the process of mediation.)

From Remediation to Mediation

The theory of originary technicity leads to one of the key propositions of our book: namely, that we have always been technical, which is another way of saying that *we have always been mediated*. This is not to suggest that terms such as "technologies" and "media" can be used interchangeably in any context. Indeed, one of the frequently unasked questions in new media studies concerns precisely the relation between media and technology. For Lev Manovich, the two seem to be the same thing. He thus defines new media as embracing "The Internet, Web sites, computer multimedia, computer games, CD-ROMs and DVD, virtual reality."[68] For Lister and coauthors, in turn, the term refers to "those methods and social practices of communication, representation, and expression that have developed using the digital, multimedia, *networked* computer and the ways that this machine is held to have transformed work in other media: from books to movies, from telephones to television."[69] Where Lister and colleagues' definition incorporates the concept of remediation and recognizes the vitality of older media, Manovich seems to shy away from such nonlinear logic. Lister and colleagues' language of *recombination* contrasts against Manovich's language of *substitution* (specifically, of media by the computer). Let us reiterate here that by raising questions for such a linear narrative of new media development—from, say, Russian avant-garde cinema to computing—we are not denying the significance of the study of media as specific objects with specific histories. It is the way in which Manovich *collapses* the distinction between different historical media, where for him "all media become new media," that raises questions for what we understand not just as the specificity of new media (that is, the recognition that the way analog music production and distribution works is qualitatively different from its digital, networked counterpart), but also as their singularity, where the latter stands for "the temporal and affective performativity of their functioning."[70] Yet, given that "the content of media is always other media" and that the process of remediation is ongoing, we need to do more to combine our knowledge of media objects with our sense of the mediating process that is continually reinventing them.

This idea leads us to suggest that "remediation" can perhaps be better thought of as "mediation," as this latter term highlights the ongoing aspect of the mediating process without circumscribing it too early either by human desire and action or by specific media. Our understanding of mediation nevertheless differs from the way this term is currently used in many academic debates. In its most frequent applications, "mediation" is a term from Marxist theory that refers to "the reconciliation of two opposing forces within a given society (i.e., the cultural and material realms, or the superstructure and base) by a mediating object."[71] The way this term is taken up in media studies is as a "mediating factor of a given culture" that takes the form of "the medium of communication itself." As Aeron Davis explains in his book *The Mediation of Power*, "mediation" is a term applied to the study of social and political processes. The following kinds of questions are typically asked in analyses of what he terms "social shaping": "how do individuals and institutions use media and communication, and, conversely, how do media and communication shape individuals and institutions? How, in other words, do individuals in their use of media, inadvertently alter their behaviors, relations and discursive practices?"[72] (Davis's own contribution to the debate lies in his study of how mediation works in "elite actors and sites,"[73] rather than in ordinary people.) The traditional thinking of mediation is therefore quite structuralist: on identifying some stable structures in society that are usually then placed in an oppositional, dialectical relationship (say, elites versus ordinary people), mediation is mobilized as a third intervening and negotiating factor. However, throughout the analysis, the very system—with its structuring elements (individuals, institutions, media)—remains firmly in place, even if some reorganization within the given structures (for example, change of individual behavior) and within the powers they are seen as yielding occurs as a result of such mediation. This rather static notion of mediation lends itself to the study of "effects" because it is premised on a set of determining assumptions around subjects, objects, and the relations between them. It is also usually rather humanist, in the sense that the key agents participating in, or undergoing, mediation are human—even if the term mediation refers to a nonhuman entity such as "labor" or "capital." Mediation is therefore primarily a tool for enacting some sociopolitical ends.

In his article "Mediatization or Mediation? Alternative Understandings of the Emergent Space of Digital Storytelling," Nick Couldry provides an overview of the use of the term "mediation" in various disciplines. He writes:

> As a term, "mediation" has a long history and multiple uses: for a very long time it has been used in education and psychology to refer to the intervening role that the process of communication plays in the making of meaning. In general sociology, the term "mediation" is used for any process of intermediation (such as money or transport). . . . Within media research, the term "mediation" can be used to refer simply to the act of transmitting something through the media.[74]

Couldry then goes on to acknowledge a somewhat cruder application of this concept in media studies since the early 1990s, whereby mediation stands for "the overall effect of media institutions existing in contemporary societies, the overall difference that media make by being there in our social world."[75] "Mediation" therefore seems to be just equivalent to "media saturation."[76] Couldry himself is more inclined to follow Roger Silverstone's definition of this term, wherein mediation "describes the *fundamentally, but unevenly, dialectical process* in which institutionalized media of communication (the press, broadcast radio and television, and increasingly the world wide web), are involved in the general circulation of symbols in social life."[77] It is clear from Couldry's definition that mediation is still very much about the workings of institutional media (be it in their mainstream or "alternative" guises): a process that for him occurs against the predefined canvas described as "the social." If for Couldry "mediation" remains an important term for grasping how media shape the social world,"[78] we are still here within the logic that carves out (or spatializes), and hence separates, entities such as "media" and "the social world" in order to analyze the relationships between them via an intermediary layer of "mediation." This is a rather static model, one that positions media as a primary term, a thing than then gets "mediated" and becomes part of a "media flow" as a result of something (interpretation, circulation, etc.).[79] In our own understanding of mediation, however, we postulate a reversal of such a static position. For us, mediation is the originary process of media emergence, with media being seen as (ongoing) stabilizations of the media flow.

In order to develop a better understanding of the relationship between mediation and media, as well as between media and technology, we should pursue further the consequences of the theory of originary technicity, starting once again with Heidegger. Rather than go along with Poster's assessment that Heidegger's theory does not have much to offer if we want to understand information technologies or new media, we want to suggest that it can provide a framework for understanding historically specific and uniquely singular media within the wider technological framework. For us, media need to be perceived as particular enactments of *tekhnē*, or as temporary "fixings" of technological and other forms of becoming. This is why it is impossible to speak about media in isolation without considering the process of mediation that enables such "fixings." By saying that the logic of technology (as well as use, investment, and so on) underpins and shapes mediation, we are trying to emphasize the forces at work in the emergence of media and in the ongoing processes of mediation. If, to return to Bolter and Grusin, remediation is the mediation of mediation, and if this process is ongoing—even if historically specific—then mediation seems to us a more apposite term, philosophically, with which we can describe the being and becoming of media. The definition of "media" as temporary "fixings" of technological as well as, say, political becoming must also incorporate the communicative aspect that the term "media" is traditionally associated with. Yet what we have in mind here is more than facilitation of a dialog or discourse between two human entities. Media "communicate" in the sense of *always remaining turned toward what is not them*, in being a delimitation of the standing reserve of technology that has to be temporarily cut off but that must never be forgotten. *Every medium thus carries within itself both the memory of mediation and the loss of mediations never to be actualized.*

The potentiality of mediation inherent both in the existing media and in the technological enframing (what Heidegger calls *Gestell*, which we can also translate as "setup") of the world points to the inherently creative character of mediation, even if this process of creation has to entail erasure, forgetting, overcoming, and at times violent transformation. If mediation is partly a technological process, then it partakes of the force of *poièsis* that Heidegger identifies in all *tekhnē*: a force that brings-forth or "presences" the world.[80] It does so by means of *physis* (nature) or through the activity of an artisan or artist. Arguably, many contemporary media successfully blur the Heideggerian distinction between "the growing things" and things completed "through the crafts and the arts": we can think here of "biomedia," which involve "the informatic recontextualization of biological components and processes," whereby "the biomolecular body is materialized as a mediation,"[81] bioart such as Stelarc's *Extra Ear: Ear on Arm*, soon to be equipped with a radio transmitter, and the Tissue Culture and Art Project's "semiliving" sculptures, artificial life, or even the social networking sections of the Internet that go under the name of Live Web (LiveJournal, Facebook, Flickr). The very process of media emergence involves creation, whereby human creative activity is accompanied (and often superseded or even contradicted) by the work of nonhuman forces. This process is also hybrid: it interweaves different entities, or rather it stabilizes, or "fixes," entities in the process of interweaving them. It is therefore not just that all media are remediated, as Bolter and Grusin have it; *mediation can also serve as a name for the dynamic essence of media*, which is always that of becoming, of bringing-forth and creation. However, in that process of ongoing mediation, with its inevitable ebbs and flows, singular stabilizations, fixes, or cuts to this process *matter*. Not only are these singular fixes or cuts responding to the wider historicocultural dynamics; they also, in their subsequent incarnations as "media," acquire a cultural significance.

The creative aspect of media flows is taken up by Scott Lash and Celia Lury in their book *Global Culture Industry: The Mediation of Things*, which also attempts to outline a different approach to mediation. Intended as an update to the Frankfurt School theories for the age of globalization, the book traces the circulation of media and art objects in the global world under two headings: "the thingification of media" and "the mediation of things." Interestingly, the authors go beyond the earlier static model of the relationship between media and their mediation in suggesting, after Adorno, that the media object "does not pre-exist its mediation" because the object of cinema, for example, "is itself constructed—coordinated, organized and integrated—in mediation, in mass movement."[82] Yet in insisting on the inherent mediation, movement, and hence vitality of contemporary media objects (the *Toy Story* movie, Swatch watches), Lash and Lury inscribe their theory in a rather linear framework of development. This vitality and movement seem to be for them signs only of the global age. Globalization here becomes a (philosophically if not materially) external impetus that changes the way the media universe is organized. Their frequent repetition of the "no longer" phrase in their book—as in media reproduction no longer allegedly being mechanical, no longer being "about" identity, standardization, or representation (claims that easily open themselves to being challenged)—closes the door to the philosophical positioning of mediation as the originary logic of media, while also raising questions as to the possible origins and source of that vitality. The underlying, although perhaps unintended, conclusion of *Global Culture Industry* is that it is only with globalization that media become truly moving, that they gain a new level of intensity, that they are "lively" at last.

The Vitality of Media

In the final section of this chapter, we take a closer look at the question of the vitality, or "lifeness," of media, without equating the latter with the period of intense capitalism or globalization the way Lash and Lury seemingly do. Engaging with the ideas concerning duration, creativity, and life developed by Bergson, we shall posit a thesis that mediation can be seen as another term for "life," for being-in and emerging-with the world. This thesis arises out of a body of work rooted in cybernetics and systems theory, whereby the media are understood primarily in terms of an ecology and a dynamic system of relations rather than as a series of discrete objects. Yet we also have reservations for some overenthusiastic applications of systemic thinking to the question of mediation. This is why a notion of "the cut"—as both a conceptual and material intervention into the "media flow" that has a cultural significance—is important for us in understanding what it means to be mediated, and in taking responsibility for this process, from within the process itself. (Chapter 3 discusses the notion of "the cut" with relation to photographic and other media practices in more detail; chapters 4, 5, and 6 develop issues of agency and responsibility with regard to mediation in its attempt to outline "an ethics of lifeness.")

A rarely cited article in mainstream media, communications, and cultural studies, "A Theory of Mediation" by Gary Gumpert and Robert Cathcart, deserves closer attention precisely due to the effort on the authors' part to outline the conceptual force of "mediation" beyond its more established meanings as a channel that carries information, an awareness of media producers, or a set of media effects, while also pointing to the inherent logic of mediation in what Lister and coauthors call "technological and biological things." Drawing on McLuhan's theories, Gumpert and Cathcart start by postulating that media become "inseparable from the human communication process"[83]—a state of events that disallows any idea of the full self-determination of the human. Indeed, for the two authors, humans can never escape "the determining ends of their technologies."[84] The determinist tenor of this latter phrase is modified by their recognition of the more complex and interlaced dynamics of agencies and forces at work in the media environment. They write, "There are always forces outside ourselves that limit or change our alternatives. Part of these forces are the technologies we have become dependent on. . . . No one today can operate apart from the influences of mediation, because our functional, cultural, social, and psychological identities are, in large part, dependent on the instrumentalities of media."[85]

Interestingly, Gumpert and Cathcart's definition of media is not limited to those media that have broadcast or mass potential. The authors are critical of the "limited concept of mediation that ignores several millennia of technological development and influence"[86]—cave paintings, clay tables, papyrus, architecture—and argue that a "theory of mediation should reconcile the current bias by stressing the development and continuity of media technology and its effects."[87] The most significant aspect of Gumpert and Cathcart's theory is their proposition that "human and media development are intertwined in a helixlike embrace."[88] They then go on to outline an intriguing parallelism between biological and technical evolution—an idea that has also been explored by thinkers such as Leroi-Gourhan, Simondon, Lem, and Stiegler but that gains a new inflection when linked with the concept of mediation. Gumpert and Cathcart write:

> To the extent that humans' mental processes allow them to make complex and varied extensions of themselves, one might see the development of media technology, in the loosest sense, as a form of biological development. We may infer that the principles of biological evolution can be used to explain media evolution. We could call this a biotechnological explanation of mediation.[89]

The literalness of this proposition, or its scientific "truth," is of less interest to us than its conceptual and rhetorical force, that is, the idea that looking at analogies between biological organisms and media in terms of their complexity, adaptability, and specialization may allow us to shift the perspective of what counts as media, or even lead to the recognition of the *poietic*, creative impulse, beyond that of human volition, in what Heidegger called "the growing things of nature as well as whatever is completed through the crafts and the arts."[90] *It is here again that our strategic distinction between media and mediation becomes important because, arguably, it is also a distinction between appearing "live" and becoming "life-like."*

We want to suggest that mediation, according to the way we understand it, gives us insight into the *lifeness*, or *vitality, of media*. By this we mean something more than just the *liveness* of media, which we know about through television studies of catastrophes and other "newsworthy" occurrences (which we discuss in chapter 2). We are referring instead to the possibility of the emergence of forms always new or potentiality to generate unprecedented connections and unexpected events. Media only intimate at lifeness through their appeal to "live coverage," the lively, flashing "look" of their animations and their representationalist aspirations aimed at closing the gap between the viewer and the screen; they ultimately foreclose on life because they reduce it to a linear and predictable set of outcomes. Consequently, they end up undermining "the vitality of evolutionism"[91] while also foreclosing in advance any true radical creativity and inventiveness that life is capable of generating. The apparent *live*ness of media is therefore only a mask for their enhanced social and economic utility. Yet if media can only appear to be "live," mediation can serve as an opening onto the underlying temporality of media, that is, its lifeness. To make sense of this latter concept, we turn to Bergson's intuitive method.

Bergson proposes a method or means of comprehending time through intuition. Intuition bridges instincts and the intellect, physical actions, and reactions with our habits of thought. Although it is predominantly affiliated with instinct, and is a biological more than a psychological tendency, intuition reconnects ordinary, scientific, or philosophical knowledge with life, where life is understood by Bergson to be synonymous with time, movement, and the process of creative evolution. Intuition therefore enables us to apprehend that which is in process. It is not *itself* equivalent to a process, but it is nevertheless a movement out of our own duration that enables us to connect with a wider one.[92] We are able to connect by recognizing our relationality with that which we perceive or observe. In as far as its central premises are those of process and relationality, intuition should be distinguished from affect, which—at least in its current derivation—frequently seems to operate more in terms of causation and mutuality.[93] Where affect is a synonym for the mutual effects of subjects and objects, minds and bodies, intuition signals their irreducibility. It is this irreducibility that for Bergson is denied in all forms of intellectual knowledge.

Our intellects tend to divide the object world, as a result of which we conceive of change as a succession of states from A to B, from youth to old age, from past (represented as "analog") to present (which gets equated with "digital"). We immobilize what is mobile, turn time into space, in order, Bergson suggests, to master it. Through measurement and prediction, we are able to act and intervene as if from outside and above. Having cut reality up into manageable parts, we then reassemble it, recreating movements that remain resolutely "false," or mere semblances of movement with no real duration. Notable examples, for Bergson, include Spenser's evolutionary theory and the "cinematograph." Bergson is opposed to habits of thought that he terms "mechanism" and "finalism,"[94] and that we might better understand as cause-and-effect determinism and teleology. They may be useful, principally by enabling us to act in a world rendered knowable and therefore controllable, but they are still misleading, and propose a view of time as space—which for Bergson is simply erroneous. Time is indivisible, continuous and unknowable, at least to the intellect. In order to "know" it, we must rediscover another, perhaps atrophied kind of knowledge, one more aligned with our instincts and by no means easy to access.

Intuitive knowledge, or contact with duration, is not (unlike affect) a given. It is achieved with difficulty, and even then it is fleeting—and thus impossible to sustain. Not to be conflated with "a feeling, an inspiration" or a "disorderly sympathy,"[95] it is perhaps more like a moment of insight that moves theory on[96] or contributes to knowledge not least by challenging its very foundations. Intuitive knowledge comes at the cost of ready-made categories and concepts. It is literally thinking "out of the box." Antithetical to knowledge-as-we-know-it, intuition opens us to the possibility of knowledge-as-it-could-be[97] and eschews generalities for specificities, representations for a certain realism toward events in process. It also implies a different mode of communication—more analogical, imagistic, metaphorical—that might seem anathema to the conventional scientist or even the more professionalized humanities scholar, but not, perhaps, to the philosopher-feminist or the artist.

Bergson's distinctions—between true and false problems, differences in degree and differences in kind—are not absolute but rather methodological. Through intuition, he seeks to distinguish between entities in order to understand the relation between them better. However, his concepts of time and space, life and matter, remain, at least in parts of his work, problematically unrelated, and this may have encouraged, in subsequent academic work inspired by his writings, a habit of thinking in terms of process without the specificity, "precision," or singularity that Bergson himself stresses in *The Creative Mind: An Introduction to Metaphysics*. We should note that in his emphasis on process, Bergson is not interested in explaining the "totality of things."[98] He is rather interested in the particular "thing" that demands not a mere restatement but rather a reinvention of the philosophical problem[99] and that involves an effort to become conscious of becoming that extends the self in the direction of the other.[100]

Given that the particular "thing" of which Bergson speaks is not fixed—at least if it is living, organic—it may be conceived of not so much as an object but as an event. As Derrida puts it in an interview titled "Artifactualities,"

> The event is another name for that which, in the thing that happens, we can neither reduce nor deny (or simply deny). It is another name for experience itself, which is always experience of the other. The event cannot be subsumed under any other concept, not even that of being. The "there is" ["*il y a*"] or the "that there is something rather than nothing" belongs, perhaps, to the experience of the event rather than to a thinking of being.[101]

Bergson also identifies the event or eventness of the thing as an experience predicated on the sacrifice of familiar concepts and categories of thought. "The truth is," he says, "that an existence can be given only in an experience."[102] He rejects "words" in favor of "things,"[103] or, more precisely, he rejects the equivalence between words and things that characterizes language as a nomenclature and representation as realism. Seeking to bypass language by referring to the *inexpressibility* of intuitive knowledge, Bergson posits intuition as the "direct vision"[104] of the mind and of life (if not matter).

Our adherence to Bergson's philosophical method does not extend to this double separation: of the event from language and mediation, and of life (understood, ultimately, in spiritual terms) from matter. Instead, we will expose Bergson's three basic laws or rules of intuition to Derrida's argument about the irreducibility of the event and its mediation, and to Barad's insistence on the dynamism of matter across all scales, down to and including, in our case, particles—the "building blocks" of life. For Derrida, it is not so much that the event can only be experienced; it is that the event constitutes experience that, in turn, is always already mediated by teletechnologies, such as television and other telecommunication networks, but also language itself. This experience does not have to be fully conscious and will always, to some extent, escape articulation. Yet, for us to try and make sense of it, to incorporate it into the theoretical framework of "mediation," we need it to become part of the work of *différance*, whereby we recognize that terms such as "media," "event," and "experience" "have not fallen from the sky fully formed, and are no more inscribed in a *topos noētos* [the eternal world of ideas], than they are prescribed in the gray matter of the brain."[105] If we take mediation as our central problem and seek to establish its difference in kind from media by emphasizing its temporality, we do not do so at the expense of singularity. Indeed, we are driven by a requirement or even an ethical injunction to cut across the flow of media in order to say something about them.[106] Doing so requires us to always consider time in relation to space, matter in relation to life. Even if the states of media are contingent actualizations of the process of mediation, the fact remains that, as Derrida has it, "there is not *only* process."[107]

Mediation is therefore also a differentiation, a "media becoming," that is always at the same time a process of "becoming other." Our book's philosophical trajectory is not therefore faithfully "Bergsonian." Instead, we aim for a critical encounter between Bergson's notion of "creative evolution" (and Delezue's embracing of its spirit) with Derrida's notion of "différance," in which there can be heard echoes of both deferral and distantiation and which suspends any absolute point of departure in the production of meaning.[108] Indeed, it is around the notion of differentiation that we part company with Bergsonism. Bergson accounts for difference as immanent differentiation from within that drives the production of "forms ever new." This idea comes head to head in this volume with the Derridean understanding of difference as a (quasi)transcendental place of absolute alterity that cannot be subsumed by the conceptual categories at our disposal because these very categories rely on the process of differentiation, on not-being-other-categories. Even if the two philosophical traditions—the "immanentist" one as encapsulated by the work of Spinoza, Bergson, and Deleuze, and one of transcendence as developed by philosophers such as Heidegger, Levinas, and Derrida—both share an interest in the possibility of change,[109] we will attempt to negotiate between "differentiation within" the flow of live, biotechnical media on the one hand, and, on the other hand, the cut or interruption to this flow of difference that comes from a position of a formal outside, and that calls us to a response and responsibility toward the other.[110] The negotiation between the Bergsonian and Derridean philosophical traditions is nevertheless of interest to us only as far as it allows us to think, move with, and respond to the multiple flows of mediation.

Derrida's term *différance*—a word that looks like a "kind of gross spelling mistake"[111] but that loses its "erroneous" status when pronounced and thus introduces both error and play into discourse—points to precisely such spatiotemporal meanderings and negotiations. "In the delineation of *différance*," writes Derrida, "everything is strategic and adventurous. Strategic because no transcendent truth present outside the field of writing can govern theologically the totality of the field."[112] Any pretense toward rooting media analysis in "the social," "subjectivity," or indeed the unproblematic notion of "media" must therefore be seen as nothing more than pretense. It is not that many traditional forms of media analysis do not recognize this pretense, or the need for it, but what we find lacking in many of them is the kind of adventurousness Derrida is talking about, where this strategy of spatiotemporal differentiation is not orientated toward a final goal or theme but is rather a "strategy without finality, what might be called blind tactics, or empirical wandering if the value of empiricism did not itself acquire its entire meanings in its opposition to philosophical responsibility."[113]

What will therefore hopefully emerge through this process of playful yet philosophically rigorous intervention in this book will be a more dynamic, networked, and engaged mode of working on, in and with "the media," where critique is always already accompanied by the work of participation and invention and where empiricism becomes a serious strategy of both "doing things with words" and "doing things with things."

2 Catastrophe "Live"

The Credit Crunch and the Really Big Crunch

This chapter examines two media events that are linked by the prospect of global or even cosmic catastrophe at the dawn of the twenty-first century: the credit crunch—a term widely used in the British media to describe the global financial woes originating with the US subprime mortgage crisis and encapsulating the European sovereign debt crisis in late 2000s—and what we will be referring to as the "big crunch," otherwise known as the Large Hadron Collider (LHC) project developed at the European Organization for Nuclear Research, CERN, in Switzerland. We will discuss the staging of the LHC project in and by the media, comparing it with the credit crunch as a historically contiguous yet generically different media event.

Such a comparative reading of the economic and the cosmic as contiguous yet different media events is justified on a number of levels. It is precisely through recourse to cosmological metaphors that Dhaval Joshi at BCA Research—one of the world's leading independent providers of global investment research—has described the European crisis over the shared euro currency in November 2011: "Approaching a black hole, cosmologists define the event horizon as the point beyond which it is impossible to escape a guaranteed ultimate annihilation," Joshi said. "The fascinating thing is you can cross this point of no return without realizing that your doom is certain. So the question is: has the euro area unwittingly crossed its own event horizon? We believe not, although it is getting dangerously close."[1] Significantly, Joshi captures the whole set of intertwined economic debates; political negotiations; falling financial indices such as FTSE 100, Dow Jones, or Nikkei and their graphic representations; and the ensuing business and personal financial losses with the all-encompassing term "event horizon," a key breakthrough moment after which nothing will ever be the same—or even nothing will ever *be*, full stop.

However, it is not just the ominous sense of something really bad happening to the universe as we know it that makes us stage this comparison between the impending economic and cosmic disaster here. It is first and foremost the construction of news items about these happenings as "media *events*" (of which Joshi's diagnosis is just one example) and the ensuing but frequently overlooked processes of dynamic material mediation involved in them that are of interest to us. With this, we seek to contrast the idea of an atypical and actually or potentially catastrophic media event (the end of the cosmic universe as we know it) with that of an ongoing and more dynamic mediated event (the current undermining of the financial security and stability, which may or may not lead to the end of the socioeconomic universe as we know it). Through generating images and forms of knowledge, such as physics and journalism, the two events we want to look at here, in addition to *describing* the credit crunch and the big crunch, may be said to contribute to bringing about these two "crunches." It is, we will argue, this more direct yet differentiated *material connection* between media and the events in question that draws our attention to problems of time as well as space, processes as well as objects, performativity as well as representation, mediation as well as media. Above all, the credit crunch and the big crunch reopen and enable us to restage the problem posed by Jean Baudrillard: that of the relation between the (catastrophic) event and its mediation.

The credit crunch is something with which most citizens of the neoliberal world have had a first- or at least second-hand experience, both through being exposed to media stories about it and through witnessing the financial and emotional transformation to social and individual lives as a result of it. A wealth of media images easily spring to mind, including Wall Street, Tokyo, and London stock exchange traders in a state of animated despair; graphics tracking the dramatic devaluation of the yen, the dollar, the pound, and the euro; employees walking out of Lehman Brothers carrying cardboard boxes; and the prime ministers of Greece and Italy, George Papandreou and Silvio Berlusconi, visibly suffering the glare of global media while the presidents, chancellors, and prime ministers of countries less close to bankruptcy strive to maintain composure. On the other hand, the big crunch—whose physical and, indeed, metaphysical implications are not any less significant—may need some further explanation. *The Observer Book of Space* defines the big crunch as follows: "If there is too much matter in the universe, its expansion will slow and finally stop due to the massive gravitational pull of all this mass. The universe will not just stop there, though, but quickly reverse, pulling all matter back into the infinitely hot, infinitely dense singularity that existed before the Big Bang."[2] The big crunch, then, is officially the opposite of the Big Bang; it is a contraction of the universe that marks the end, rather than the beginning, of life as we know it. Although the LHC project aims to recreate the Big Bang and the conditions that prevailed immediately after it, the prospect of a crunch has arisen around the possibility—an entirely imaginary one—that the experiment at the heart of the project will cause black holes to form and these black holes will subsequently absorb, or swallow up by gravitational means, the entire universe.

By reviewing the trajectory of debates on media events in this chapter, we will suggest that the LHC project appears to conform to the outmoded ceremonial tradition originally defined by Daniel Dayan and Elihu Katz in *Media Events: The Live Broadcasting of History*, whereas the credit crunch might be seen as a key example of the disruptive news event, which has increasingly "upstaged" the early television ceremonies of contests, conquests, and coronations. However, we will also offer a critique of debates on media events in as far as they rely on Émile Durkheim's religious sociology and on "a Cartesian habit of mind" that creates and utilizes relatively static and homogeneous categories such as "media" and "society."[3] Instead of thinking in terms of categories of things that can have effects on other things—effects such as unification or fragmentation, celebration or commemoration—we will turn to an anti-Cartesian philosophy that presumes an a priori connection between such "things." In other words, we will argue that media cannot have *effects* on society if they are considered to be always already social. From this perspective, the questions we can ask about the media events and their effects change from whether, or to what extent, media events integrate (or *dis*integrate) society—as if the latter were something separate, simply existing *out there*—to how media produce or enact the social. We will show how this performative rather than representational aspect of the media is indicated but generally not pursued in mainstream studies of media events.

William Merrin suggests that research on media events has proliferated as a result of "developments in the technology, scale, spectacle, ubiquity and realism of media coverage" as well as "changes in broadcasting structures."[4] This increasing interest in media events—in their liveness, ritual, and spectacle—has in turn resulted in multiple analyses of specific events and in "the development of new interpretive approaches."[5] *We want to suggest that understanding media events in terms of their mediation can be seen as one such new interpretive approach that has emerged within the study of new media in general*[6] and in discussions of media events in particular.[7] We seek to develop this notion of mediation here as part of our critique of representationalism. For Karen Barad, representationalism is characterized by "the belief in the ontological distinction between representations and that which they purport to represent [whereby] . . . that which is represented is held to be independent of all practices of representation."[8] In

Creative Evolution, Bergson places representationalism at the heart of scientific, ordinary, cinematic, and photographic knowledge. He argues that it produces useful ideas that are ultimately false or misleading because they are predicated on ontological distinctions and spatializations that do not actually exist.[9] In *Simulations*, Jean Baudrillard seeks to shatter the belief in such ontological distinctions and highlight the death or disappearance of the "representational imaginary,"[10] and of the real within simulation. Our aim here is to distinguish between simulation and mediation and to show how the latter signals the performative aspect of media, whereas the former, in contrast, frames media in terms of the negation of events.[11] With reference to our intertwined account of the credit crunch and the big crunch, we want to investigate the extent to which media bring about, or contribute to bringing about, the event that they subsequently describe. What is of particular interest to us here, then, is a productive, rather than negative, relationship between the event and its mediation.

The question of the relation between the event and its mediation[12] is arguably central in the context of the credit crunch. In the United Kingdom, however, it was eclipsed by another question: did Robert Peston (BBC Business Editor) *cause* the recession? A question pertaining to media relationality has thus been superseded by one pertaining to media causality. Although the question of relationality is also arguably the key question in the context of the LHC project, it has been displaced, worldwide, by another: will scientists *discover* the Higgs boson, otherwise known as "the God particle"? Both of these superseding questions about causation and discovery have a representational premise. They assume the existence of representations (not just images, but forms of knowledge such as physics or journalism) on the one hand, and of events such as economic crises or objects such as particles that are supposedly able to be represented/misrepresented and found/lost, and that are subject to the principle of cause/effect, on the other. In what follows, we shall seek to expose the limits of this assumption and pursue the implications of performativity for the study of media events.

From Media Events to Mediation Processes

According to Dayan and Katz, "Media events are events narrated . . . by television . . . [yet] their origin is not in the secular routines of the media but in the 'sacred center' (Shils 1975) that endows them with the authority to preempt our time and attention."[13] Nick Couldry explains further that the two theorists define media events "as preplanned, but non-routine, live transmissions of real events . . . [which] do not just relay what would have gone on without them, but rearticulate the elements and sites of an existing ritual process into a fully *mediated* event whose form was unimaginable before electronic media."[14] It has to be acknowledged that Dayan and Katz did not initiate research on media events; such studies date back to the 1950s, when Shils and Young published their analysis of the Queen's coronation.[15] Nevertheless, Dayan and Katz's work is a key reference point in media and communications studies. Interestingly, it is treated in much the same way as the two authors treated their subject matter: it is defended, though not uncritically. In their seminal work of 1992, *Media Events: The Live Broadcasting of History*, Dayan and Katz set out to defend the phenomenon of the media event from the accusation, notably by Daniel Boorstin,[16] that a media event is only ever a pseudo-event, a depoliticizing spectacle that services the operations of power by masking them. Dayan and Katz clearly disagree with this position. They ask: "When 87 percent of Israelis sit at attention in front of their television sets to view the opening ceremonies of Holocaust Day . . . are they victims of a yearning for unity—for that is what it is—that is undermining their parliamentary democracy? . . . Perhaps so, but we think otherwise."[17] The two authors seek to make sense of their own experience of enchantment evoked by the magic of events such as the signing of a peace accord between Egypt and Israel, the marriage of Charles and Diana, and the first human steps on the moon. In order to articulate a sense of occasion that is *transcendent*, out of the ordinary, and at the same time shared with the nation, if not the world, they turn to Durkheim's religious sociology and, specifically, to his division between the sacred and the profane. Sacred rituals interrupt routines and function to unify individuals. They are forms of socialization for Durkheim.[18] For Dayan and Katz, media events can be best understood as sacred rituals, ceremonies, or festivals. The authors refer in particular to the "festive viewing of television" and to the "invitation" or even "command" that people "stop their daily routines and join in a holiday experience."[19]

Ceremonial media events are exceptional events for Dayan and Katz: they monopolize air time; they happen live; they are organized, courtly, and awe-inspiring.[20] Significantly, they differ from news events by shifting the focus from conflict to reconciliation. They "call for a cessation of hostilities, at least for a moment," and focus on heroic acts and establishment initiatives "that are therefore unquestionably *hegemonic*."[21] Media events and news events are different genres of broadcasting for Dayan and Katz, genres that represent, respectively, the integration and disintegration of society. Media events "restore order."[22] They are, indicatively, the Kennedy funeral rather than the Kennedy assassination.

Yet in a recent review of their earlier work, Dayan questions the extent to which television may still be regarded as "the medium of national integration."[23] Multichannel

TV, the Internet, and mobile communication devices effectively segment and subdivide audiences, resulting in the state of events when "television-as-we-knew-it continues to disappear," with no obvious replacement.[24] Changes in television and in the configuration of nation states could be seen to undermine the unifying premise of their media events thesis.[25] However, Dayan refers to his and Katz's original observation that "the genre of media events may itself be seen as a response to the integrative needs of national and, increasingly, international communities and organizations."[26] Putting aside for the moment the question, or rather presumption, of "integrative needs," Dayan is raising the possibility that media events become more, not less, relevant in the context of technological development and globalization. He rehearses the main stories—contests, conquests, coronations—and rephrases the main characteristics of media events he set out, with Katz, in 1992:

Insistence and emphasis
An explicitly "performative," gestural dimension
Loyalty to the event's self-definition
Access to shared viewing experience.[27]

In his own retrospective, co-authored with Tamar Liebes, Katz reminds us that conquests are those "great steps for mankind," moon landing events, and that contests include sports and political events such as the Olympics and presidential debates.[28] Coronations may include weddings and funerals: "the role-changes of the mighty."[29] He and Liebes list eight characteristics that they see as common to all three narratives: "(1) the live broadcast, (2) the interruption of everyday life and everyday broadcasting, (3) the pre-planned and scripted character of the event, (4) the huge audience—the whole world watching, (5) the normative expectation that viewing is obligatory, (6) the reverent, awe-filled character of the narration, (7) the function of the event as integrative of society, and typically, (8) conciliatory."[30]

The Ultimate Conquest?

Drawing on Katz and Liebes's categorization, we can suggest that the LHC project offers the ultimate conquest narrative. It is Big Science on a cosmic scale. The project seeks the very frontier of scientific knowledge and turns particle physicists—one in particular, as we shall see—into amiable heroes who will explain, as God himself might, the origin of life, of the universe, and of everything else. If, as Hepp and Couldry suggest, there is always an element of discovery and suspense in a media event narrative, an element of not knowing what will happen and whether the hero[es] will succeed, then this element is multiplied exponentially in the LHC project, which aims to find out what, exactly, existed at the origin of the universe and, at least in the popular imaginary, risks *everything* in the attempt.[31] The quest for the so-called God particle is effectively a quest for ultimate knowledge. The symbolic stakes are so high that they seem to involve salvation and damnation rather than mere success and failure. Here, the conquest narrative is revealed as archetypal and Promethean. It is concerned with the possibility that humankind is significantly overreaching itself, and with consequences that—thanks to the narratives presented in *Doctor Faustus, Frankenstein, Paradise Lost,* or the Bible—are already well known.

The narrative aspect of the LHC project was highlighted, directly and indirectly, in its publicity. This narrative very much shaped the launch of the project, which took place on September 10, 2008. The launch, or switch-on, of the LHC constituted a media event, according to the criteria set out by Dayan and Katz. It was preplanned and broadcast live, and it interrupted the normal programming schedule when the BBC devoted not just the launch day itself—the "Big Bang Day"—but a whole week to the coverage. The United Kingdom is a major investor in the LHC, which is said to have cost in the region of £5 billion, but the project is a multinational endeavor involving some thirty-eight countries and was broadcast widely during the launch. Although Hepp and Couldry have questioned the extent to which the concept of a media event remains tenable in a global context,[32] the LHC certainly stakes a claim to being a global media event. Rhetorically, it functions to integrate "us" at the highest (or lowest) possible scale, not as a nation or even a species (the human race) but rather as a constituent in the universe of particulate matter.[33] Conciliation is achieved by incorporating the United States, a previous competitor with the European Union, as a collaborator in the construction of what is now the world's largest supercollider. More than that, conciliation is effected through the metaphor of people as particles—ubiquitous, equivalent, alike, and elemental. In the context of ongoing globalization, decentralization, and deterritorialization, "we" are brought together again, reformed as the undifferentiated elements of a proto-universe.[34] History, geography, politics, society, and identity are effectively and symbolically erased in order to be started afresh, renewed, and perhaps also redeemed. The "loss of the 'we'" that is brought about, according to Dayan, by the upstaging or "banalization" of media events by news, is recovered in this context, and the effect is the opposite of "disenchantment."[35] Anthropomorphized particles, as we shall see, can be very enchanting indeed.

Dayan admits that more recently he and Katz have each been "contemplating whether there is a retreat from the genres of media events, as we described them, and an increase in the live broadcasting of disruptive events of disaster, terror, and armed conflict."[36] Without having to speculate about the integrative needs of a global community, it would appear that the LHC project offers evidence to the contrary and in fact reinforces their earlier suggestion that the genre of media events is in some way a *response* to the formation of "communities larger than nations."[37] If we characterize this response as one of reterritorialization, articulated, as Dayan and Katz originally pointed out, in media beyond the reach of television, then we must also account for the ongoing *dynamic* of de- and reterritorialization. This dynamic is a facet of media understood as *processes* of mediation, rather than as forms of representation.

Just More Bad News

Dayan and Katz recognize that media are not just forms of representation. Their book was inspired by a sense that media events are actually performative and "have nothing to do with balance, neutrality, or objectivity."[38] This sense, however, does not seem to be very well developed. For them, media "actively create realities" rather than describe things.[39] Yet there is a suggestion that these realities are alternative ones, taking place in the separate sphere of the media that subsequently *have effects*—good or bad, unifying or fragmenting—on the world "out there." If news events are indeed "disruptive," then there has to be an object—namely society—that is disrupted by them.[40] This presumes a degree of separation between events and their mediation, and such presumptions are ultimately associated with representationalist thinking. Indeed, it is the apparent decline of representationalism, the closing of the gap between events and their mediation, that puts at risk, for Dayan (and for Baudrillard), "a certain form of 'enchantment,'" and creates in its place a "gray zone" and even a "simulacrum."[41]

It is not just "bad effects" that are at fault here, then, but also seemingly "bad content." Katz and Liebes argue that "live broadcasting of disruptive events such as Disaster, Terror and War are taking center stage" and upstaging media ceremonies.[42] Although such disruptions to the status quo should, retrospectively, have counted as media events, they play a different role and have different effects, leading to the creation of more scattered and more cynical audiences with less faith in government and "Great Men."[43] News events also invade ceremonial events that "may suddenly yield to an unplanned disruption."[44] Indeed, we could argue that part of the excitement of the LHC project is precisely the possibility that "something may go wrong."[45] The relentless rise of traumatic events has been accompanied, Katz and Liebes suggest, by a corresponding rise in "obsessive coverage" and even "disaster marathons" that linger on a single catastrophe for hours, days, or more.[46] In the context of terrorism, Katz and Liebes explain the symbiotic relationship between events and their mediation in terms of publicity and the control of the process of mediation by antiestablishment figures:

> Media events, of the ceremonial kind, are essentially co-produced by broadcasters and organizers such as the International Olympics Committee, the League of Women Voters, and the Royal Family. They are establishment events, with wide public support, based on mutual agreement as to how the event will be staged. Disaster marathons, on the other hand, are obvious threats to establishments, in which the organizers—the perpetrators—are an invasive force, far out of the reach of establishment control.[47]

Here, the news event is regarded as a reflection of hegemonic power, a means of making present the struggle between establishment and antiestablishment groups. The previously implied idea of performativity gives way to a rather naïve account of representation understood as transparency, immediacy, and innocence. In the wider field of debates concerning media events, performativity is sometimes conflated with constructionism or with a nontransparent, noninnocent notion of representation that is always a re-presentation. For Hepp and Couldry, for example, performativity refers to "the constructive character of media events" and is tied "to struggles for power and influence."[48]

Couldry also critiques the notion of the centering, integrating sacred ritual as envisioned by Durkheim and used by Dayan and Katz to describe the function of media events. He follows Bloch and Bourdieu in arguing that ritual, including media ritual, is not "the affirmation of what we share" but rather a means of managing conflict and masking social inequality.[49] Media events as rituals are not transcendent or sacred for him; they are essentially secular and "tied to modern forms of government."[50] Although it demythologizes any sense or experience of human togetherness, exposing it as a cover for social division, Couldry's argument restricts the potential scope of media performativity by conflating it with constructionism. He writes: "Media events are . . . constructions, not expressions, of 'the social order,' processes which construct not only our sense of a social 'center,' but also the media's privileged relation to that 'centre.' Media events, then, are privileged moments, not because they reveal society's underlying solidarity, but because they reveal the mythical constructs of the mediated center at its most intense."[51]

The problem with constructionism is that it relies on the same theoretical presumptions and category divisions as representationalism. Media, as instruments of power, are said to construct the event as an idea, in this case, an idea of "the center."[52] This *idea* of the center masks the *reality* of inequality and conflict.[53] But what if the role of media is not only to perpetuate ideas but also to bring about realities? What if, as Bergson suggests in *The Creative Mind*, the division between the ideal and the real does not hold? This is exactly the question raised by the concept of performativity. The theory of performativity—drawn here from the work of Austin, Derrida, Butler, and

Barad[54]—allows us to recognize, in short, that media are *generative*, that is, that they are *part of* the material world and do not thus exist *apart from* it. Neither a reflection of nor a mask for the social, media actively contribute to the production of the social. In other words, media perform the social—sometimes alongside and sometimes in conflict with other agencies that are not solely establishment or antiestablishment.

For Joost Van Loon, media are transformative more than they are [mis]informative. They "generate terror, they externalize our neural systems, they drain confidence, they perform 'the crash.'"[55] He argues, as we will, that the credit crunch is therefore a performative media event, meaning that it was, to an extent, generated *by the media*. Although the credit crunch is not a single event like a terrorist attack or a natural disaster, "it does consist of a series of related events: record-breaking dips in the value of shares, banks and mortgage lenders filing for bankruptcy, emergency speeches, votes in parliaments, resignations, all play their part."[56] These events do not exist outside of their mediation, which, to varying degrees, has been recognized within the field. However, this statement raises some more fundamental questions that have so far been elided, such as: "What is mediation?" and "What is the relation between events and mediation?" These are quintessentially philosophical questions, which may explain why they have been avoided in a debate that is primarily sociological. Couldry, for example, rejects philosophy as a framework for thinking about the media on the grounds that it is too abstract and general[57]—even though his own argument relies on abstract and general categories such as media, society, and power. Van Loon, in contrast, embraces Heidegger's existentialist philosophy in order to argue that media "have opened up existential moments in which we can engage with the question of being mediated."[58] Merrin argues that the value of Baudrillard's philosophy of the event has not been sufficiently acknowledged in media studies, even though (or because) it offers a radical reformulation of the media. He writes: "Baudrillard reveals a mediatic operation extending beyond the mere production or inflation of 'news'; one encompassing and implicating our entire epistemological relationship to the wider world."[59] In as far as media events are nonevents for Baudrillard—those, such as the first Gulf War, that "will not take place" and "did not take place"[60]—his position could be seen as offering a challenge to Dayan and Katz on their use of Durkheim, taking them back to Boorstin.

Media Events as Nonevents

Merrin traces the chronology of Baudrillard's work on events from those of May 1968 through terrorist events to reality TV, which he sees as "a spectacular version . . . of the transformation of life itself, of everyday life, into virtual reality."[61] For Baudrillard, as for Boorstin, all events are nonevents because they are substituted by signs, modeled and transformed into simulations by a voracious semiotic media. The "substituting of signs of the real for the real itself" is "an operation to deter every real process."[62] The result is that "all events are absorbed."[63] Following McLuhan, Baudrillard maintains that the content of media is only other media that refer to themselves in a closed, imploded circuit of hyperreality. He wants to "get rid" of the representationalist idea that media distort reality, because for him they absorb all distinctions, including those between accuracy and distortion, truth and fiction.[64] Along with the real and the event, meaning itself "manages to disappear in the horizon of communication. The media are simply the locus of this disappearance."[65] If simulations result from the disappearance, or, more specifically, "implosion" of events and their meaning in media,[66] then, as Merrin argues, "Baudrillard recasts McLuhan's 'implosion' not as a path to the real but as a process operating on and coalescing with it to produce hyperreality: a 'real' consumed in the comfort, distance, and security of the sign, giving a vicarious '*alibi* of participation' in a world whose semiotic actualization and dramatization represent instead a systematic distantiation."[67] So, far from reconstituting the social by means of sacred rituals, ceremonies and festivals, media for Baudrillard institute a "*waning of the sacred*."[68] Media events are therefore "only a continuation of the profane by identical means."[69]

Baudrillard contradicts Dayan and Katz's neo-Durkheimian reading of media events, although he remains himself a neo-Durkheimian, invested in the *collapsed* dichotomy of the sacred and profane, which he aligns with the symbolic and the semiotic, the real and the virtual, respectively. Much of his work reads like an elegy for the real and for representation, and even if it aims to be more strategic than this, constantly seeking to overturn the simulacrum and to contribute to its own implosion, it is still marked by a certain fatalism, an exaggerated and ultimately misguided investment in loss. Part of what is lost, Baudrillard argues, is the possibility of a real catastrophe. Why? "Because we live under the sign of *virtual* catastrophe."[70] The Wall Street crash of 1987 was, he argues, just one in a line of nonevents, of "'bombs' that don't go off."[71] Whether we are faced with a "financial crash," a "nuclear showdown," or a "population time bomb," the fact is "we experience no such explosions."[72] Although a real economic crisis did take place in 1929 "and Hiroshima really happened," war and money are now "hyper-realized"; they exist in "a space that is inaccessible." No matter what actually takes place, this hyperreality and inaccessibility leave "the world just as it is."[73] In hindsight, we might recognize Baudrillard's "imaginary economy" in which debt has gone into orbit, "circulating from one bank to another, or from one country to another, as it is bought and sold," in the post–credit crunch milieu of the early twenty-first century.[74] But we might also recognize that this imaginary economy was and is translatable into "real economic terms," and that the world is in no sense left intact, having experienced what, for many, is a "true catastrophe," a very real financial crisis.

Baudrillard's work is valuable insofar as it questions the relation between media and events, refusing to accept what representationalism presumes—namely, their a priori separation. However, it is also epistemologically flawed because, in exchanging separation for implosion or collapse, it still posits an outside, a real world out there that is effectively lost, substituted by simulations that absorb events like black holes absorb light.

Mediation, Not Simulation

The concept of simulation is founded on, and founders on, the collapsing of Cartesian divisions such as the one between the virtual and the real. For Baudrillard, there is simply "no difference" between the two.[75] The concept of mediation, in turn, retains such a difference, albeit with reference to the relationality rather than autonomy of media and events, of the virtual and the real. Mediation is positioned within a non-Cartesian framework (rather than an inverted or collapsed Cartesian framework) in which the virtual does not absorb and negate the real. The virtual does not substitute the real but rather produces it. Van Loon argues that mediation is a productive rather than representational process, a transformative rather than informative one, and that it "orders the world by calling it into being."[76] Mediation is disclosed in media events that open up "existential moments" of awareness, of our awareness of being in the world. Media events are thus able to disclose "being-as-mediated" by virtue of anomaly, of "standing out in time."[77]

Though he certainly offers a dynamic view of mediation, Van Loon accepts that "the essence of technology is revealing" and argues that mediation is "a technological practice."[78] He is arguably too tied here to Heidegger's thinking on technology, as a result of which he effectively conflates mediation and technology, reducing the former to a single agential force that does not so much incorporate "us" but that is rather revealed to, or concealed from, "us" as more or less self-contained individuals, that is, beings in the (ultimately separate) world. Against this, we will argue that mediation is a multiagential force that incorporates humans and machines, technologies and users, in an ongoing process of becoming-with that is neither revealed nor concealed but rather apprehended intuitively—inevitably from inside the process. Bergson's philosophical method, with his dynamic ontology of becoming-with, can serve here as a useful rejoinder to Heidegger's humanist and individualist phenomenology of being-in. The difference between the two philosophical positions, as suggested by Rosi Braidotti, has to do with understanding the opposition between relationality and autonomy and the degree to which we regard ourselves as either a work in progress, connected to other works in progress, or as the finished entity.[79]

In contradistinction to the dominant position in media studies, we will also argue that media events are not anomalous but rather heightened phenomena, meaning that, grasped intuitively, they do not constitute "time out" from "the ordinary everydayness" of mediation—that is thereby revealed.[80] Instead, for us such events constitute "time in" or the opportunity to recognize a process of which "we" are a part.

The Credit Crunch: Mediating the Economy

Following Dayan, and Katz and Liebes, it is possible to categorize the credit crunch as a news event, or an ongoing series of news events, marked by extensive if not "obsessive coverage" and a disastrous and disruptive phenomenon, namely the crunch itself.[81] However, as discussed previously, for us the problem with this approach lies in the way in which it curtails the active role of media, relying as it does on the dichotomy of media and society (in which one can *have effects* on the other) and thus, ultimately, retreating to a form of representationalism in which all that media produce is knowledge that is being deemed either good or bad, more accurate or less accurate. Indeed, what stands out about the credit crunch for us and what makes it a "media event with a difference," so to speak, is the degree to which it highlights the role of media and mediation, exposing the tension between performative and representational accounts of all media events. If this tension is held in the deliberately provocative and rhetorical question of whether Robert Peston, BBC Business Editor, caused the recession in the United Kingdom, then it is relieved by the creed of journalism and journalists, not least Peston himself—who continues to reaffirm his faith in his own objectivity.

Although Peston is regarded as the "face of the credit crunch" in the British media,[82] we do not intend to personalize or render parochial a global economic event that has roots in subprime mortgage lending and the repackaging of debt.[83] Rather, by focusing on Peston's *announcement* of the crisis at the Northern Rock bank and the way that this is said to have *precipitated* a crisis at Northern Rock and in the banking sector more widely, we aim to provide a specific case study in the mediation of an event—a case study in which the meaning of mediation is contested yet ultimately closed down.

On February 7, 2009, Singaband posted on YouTube their "proper old-timey" musical tribute to Peston and the Buchanan Brothers ("(You'd Better Pray to the Lord) When You See That Robert Peston"). Here, Peston's own creed of journalistic purity and invisibility is challenged, and he is regarded as a terrifyingly visible sign of the apocalypse in which we will be held to account, collectively, for "the trouble that's about." We might note at this point a refreshing change of tune. Although Peston has been accused of causing the recession, he himself has sought to shift the blame back on to the Northern Rock bank, with its allegedly reckless, greedy business strategy.[84] Yet in a post-crunch culture, Peston has never been the sole focus of blame or abdicated responsibility. Individual banks (including Goldman Sachs in the United States),[85] bankers, economists, regulators (including Bob Diamond of Barclays and Alan Greenspan as chair of the Federal Reserve), and politicians (including Gordon Brown as Britain's Chancellor from 1997 to 2007, and then Prime Minister from 2007 to 2010) have all ducked and pointed in the line of fire. But, as Alex Brummer suggests, many people have played the housing market and benefited from the boom—through cheap loans and widely available mortgages—at least as much as they have suffered from the bust. In this sense, as citizens of the wealthy neoliberal democracies we are all part of a culture of individual gain, one that lacks "self-control" and that cuts across "differences in income, race and geography."[86] Others, including Robert Peston, would seek to confine culpability to the West, and here in the West we are at least faced with the *prospect* of having to "mend our ways," as Singaband suggest. This prospect is made more palatable by the usually unsung suggestion that "Judgment Day" might therefore be postponed.

When it comes to declaring disaster, catastrophe, or even apocalypse, the chosen medium has tended to be television—at least until Dayan and Katz announced in 1992 that, along with the nation-state, TV was probably on its way out.[87] Television is the medium that shows us events as they are happening. For some, its very essence, or ontology, is liveness; others are quick to critique liveness and immediacy on television as a construct and hence an illusion.[88] Derrida, for example, argues that there is no liveness on TV, "but only a live effect [*un effet de direct*], an allegation of 'live.'"[89] Whatever its effects, for Derrida any given broadcast "negotiates with choices, with framing, with selectivity."[90] Television for him is thus associated with "artifactuality," or the *production* of liveness as an artifact.[91] Whatever is happening, the very temporality of a happening is "calculated, constrained, 'formatted,' 'initialized' by media apparatus."[92] Actuality, as it comes to us on TV, is therefore a fiction, "no matter how singular, irreducible, stubborn, distressing or tragic the 'reality' to which it refers."[93] However, our knowledge of television's fictitious status should not, Derrida insists, "be used as an alibi," an excuse for pretending that nothing is happening, and that reality has been subsumed in "the simulacrum."[94] Indirectly, he goes so far as to accuse Baudrillard of *denying the event*, an act which for Derrida is unethical in that it is tantamount to denying alterity.[95] In becoming aware of artifactuality, we should not lose sight of the event—which is "another name for experience itself, which is always experience of the other."[96] Derrida retains the hope "that artifactuality, as artificial and manipulative as it may be, will surrender or yield to the coming of what comes, to the event that bears it and toward which it is borne. And to which it will bear witness, even if only despite itself."[97]

The possibility of the event revealing its own alterity, which comes with the unfolding of the other through time, his or her becoming other and differentiation, situates the event outside the representational framework. Representation would inevitably amount to foreclosing the other by subsuming him or her under the conceptual categories possessed by the self—in our case, the TV viewers and the cinema audiences. Significantly, Bergson's critique of representation, outlined in his *Creative Evolution* with regard to cinema and photography, is even less compromising. It is devoid of any hope or expectation that representation might be borne toward, or bear witness to, the reality of what happens. Simply, for Bergson, time itself—also understood as duration, movement, creative evolution, and life—is what happens, but *time cannot be represented*. Any act or attempt at representation transforms time into space by cutting into the flow of reality and turning something that moves into something that is still (a photograph), or that only looks like it is moving (a film). Cinema, for Bergson, is "false" movement, an illusion, and what goes for cinematic knowledge also applies to ordinary and scientific knowledge. At best, this produces liveness, while life itself cannot be seen or known to the intellect, but only intuited: "If it is a question of movement, all the intelligence retains is a series of positions: first one point reached, then another, then still another."[98] Bergson's critique of representation is tied to his critique of the intellect, a faculty or tendency that depends on sight. Intuition—a tendency more allied to instinct, and to our primitive, animalistic connection with, rather than control of, the outside world—bypasses sight and depends rather on a moment, an instant of insight in which the temporality of the event, its happening, is properly apprehended by virtue of being experienced. Experience for Bergson does not stand for the recognition of our difference from the other but rather for connection with the (event as) other. To read time, and thus consequently duration, movement, creative evolution, and life as something present in any medium is in a way to read Bergson against Bergson. For him, presence is the impossibility of representation. There is, so to speak, no life on TV, only the effect of life—which, as Laurie Anderson sings, "tones it down."[99]

By bringing together Derrida and Bergson here in this way, we are attempting to trouble the representationalist accounts of the media that do not take time seriously enough and that tend to reduce it to a function of space, as a result of which representations are treated as graspable discrete entities from which alterity, duration, and hence life—in both its metaphorical and material sense—have been eliminated. We are also hoping to provide a different framework for understanding the potential liveness of media events, and for responding to it responsibly, in a way that does not presume to know its nature all too early. In television studies, liveness is regarded as a construct and an ideology. "To equate 'live' television with 'real life,'" Jane Feuer states, "is to ignore all those determinations standing between 'the event' and our perception of it—technology and institutions, to mention two."[100] Yet questions arise for us as to whether technology (or the institution) can indeed be perceived as a discrete something—a layer—that exists between events and our perception of them. Insofar as we argue that television and other media do connect with what we can call, in shorthand, "real life," it is not only through constructing an idea of it but also through performing it or, in Heideggerian terms, calling it forth. Television, technology, institutions, and viewers are all agents of life that exists not *in* television, or in any other representational medium, but rather *as* the process of mediation. In Bergsonian terms, the event happens *as time*, which is why it resists representation, because to represent it means having to alter it—cut it up in order to fit it into space.

Television cannot therefore capture the live events with which it has become synonymous, as such events remain in excess of it. The events that are worthy of being preceded by the adjective "live" are, indicatively and explicitly, catastrophic events, hyperbolic happenings that interrupt the sequence of other, more mundane ones and that demand our attention *right now*. Mary Ann Doane argues that television deals "with the potential trauma and explosiveness of the present" and that catastrophe is "the ultimate drama of the instantaneous."[101] Capturing large-scale disaster, technological collapse, a failure of prediction or of progress—"the escalating technological desire to conquer nature"—catastrophic events mark an unplanned, unexpected discontinuity in "an otherwise continuous system" of representation.[102] The *Challenger* space shuttle explosion of 1986 is a frequently cited example of such paradigmatic live or news events,[103] with the terrorist events of 9/11 more recently joining the pantheon.[104] These news events interrupt the normal television schedule, are endlessly repeated, and invariably involve the direct address of an anchor live in front of the camera.[105]

A number of media studies scholars have shifted their focus in their study of media events from trying to expose the illusion of the liveness of such events to investigating their broader ontological premises and effectivities. Mimi White, for example, attempts to reassess media events' relation to catastrophe when she writes: "There are many versions of live television that have much less to do with catastrophe than with banality, space, and what I formulate as a 'television of attractions.'"[106] Douglas Kellner, drawing on the work of Guy Debord, has spoken of "television's proclivities for spatial display" in terms of the "spectacle."[107] To look at the earlier example of the *Challenger* explosion from the perspective of such current media theories, we could argue that it can be read as a tragic spectacle whose indexicality, or "having-taken-place-ness," was beyond all doubt. Its temporality, or the moment of its constitution *as a catastrophe*, was on the other hand harder to determine. Constance Penley notes that "the astronauts did not die at the moment of the explosion, but when their capsule splashed down minutes later."[108] Even as it exploded, onlookers were not sure what they were actually seeing. The event "could be definitively designated a catastrophe retrospectively, not only after it occurred, but also after sufficient time elapsed to allow for expert analysis to certify it as such."[109] The event of the *Challenger* explosion was, in fact, historical, not live; it was located in the past more than the present. The live event as such may therefore be neither represented nor representable. In this particular case, the "fatal event was not witnessed by cameras, reporters or television viewers, and was barely discussed in the 'live' reporting following the explosion."[110] Moreover, in case of any such events occurring "live," most of what is being actually covered on TV at the time of their unfolding is resolutely banal. White quotes anchor Dan Rather, referring to himself as "a newsman vamping for time" when he had to improvise in the absence of any event actually unfolding.[111] This vacuum in the coverage of a catastrophe is then filled with a retrospective or historical narrative such as, in the case of the *Challenger* explosion, the origin of the mission, background stories about the astronauts, and so on. This result leads White to complain that "the emphasis on the temporality of liveness on television (immediacy, interruption) distracts from consideration of the medium's spatial articulations."[112] Here, she directs us to the specific aspects of the spectacle, its visual attraction as well as "television's ability to *show* things."[113] In case of televised catastrophes, the available, if scanty, footage—of rockets exploding, of buildings collapsing, of smoke billowing—is endlessly replayed, and "relatively static images are shown for a considerable duration, often with little or no variation."[114]

We want to suggest that the credit crunch can be usefully situated within the analytical framework discussed earlier. Despite being perhaps less dramatic in terms of its televisual effects, the credit crunch was arguably a media event, as its sudden occurrence and hence newsworthiness was marked for many commentators—not least Her

Majesty the Queen[115]—by the failure of economic prediction. The actual event of the credit crunch, its point of origin in the US subprime mortgage market, is widely debated but also contested and, in historical terms, vague. Brummer argues that "the sub-prime debacle had its origins in the US financial policies that went back to the 1990s and the low-interest rate regime that had been established by Alan Greenspan, chairman of the Federal Reserve" and head of the US central bank from 1987 to 2006.[116] The credit crunch, according to Brummer, was attributable to an economic policy. It was not, therefore, representable as an event. However, in the United Kingdom, it certainly *appeared* to unfold live on television when the above-mentioned Robert Peston made his by now infamous announcement on *News 24* at 8:30 p.m. on Thursday, September 13, 2007. That evening, having already posted his blog, Peston revealed that the Northern Rock bank had approached the Bank of England for emergency financial support. He said: "Northern Rock has been one of the fastest growing British banks, but it is now set to become famous for reasons it would rather keep quiet. It's become the first bank in years to seek emergency funding from the Bank of England in its role as the lender of last resort."[117] The good news, he added, was that the Bank of England had agreed to support Northern Rock because it regarded the bank's problem as exceptional and, it hoped, short-lived. Again, on the same evening, the BBC website confirmed that "the Bank of England has agreed to give emergency financial support to the Northern Rock, one of the UK's largest mortgage lenders." Peston is cited as declaring that "the fact the Bank is willing to act should be reassuring." He is also reported to have said that "although the firm [Northern Rock] remains profitable, the fact that it has had to go cap in hand to the Bank is the most tangible sign that the crisis in financial markets is spilling over into businesses that touch most of our lives." The crisis in financial markets stemmed from the "losses made by investors in loans to US homebuyers with poor credit history." This made them "wary of buying all mortgage debt, including Northern Rock's." All banks were having difficulties, Peston admitted, but since it had become such a specialist in mortgages, accounting for one in five of all UK mortgages, "no-one really wants to lend to Northern Rock."[118]

Questions arise with regard to the narrative construction of this report. British journalist and economic commentator Alex Brummer queries how and why the Rock's predicament had been leaked to Peston, "a reporter with unusually close links to Downing Street."[119] But, he adds, "whatever the source, the release of the information turned out to be a disaster for all those involved" because "far from calming anyone, Peston's delivery, oddly drawling and breathless at the same time, gave rise to panic."[120] It is worth pointing out here that Brummer is Peston's immediate rival and counterpart at the right-wing newspaper *The Daily Mail*, having himself written about the risks of "securitization"[121] and the issue at Northern Rock. Let us take a closer look at the comments he made in the aftermath of Peston's report:

> The BBC's business editor had a brilliant scoop, but the normally cautious broadcasting network had inadvertently *precipitated a crisis*.
>
> Pandemonium broke out on the morning of 14 September. . . . Fearing that they could lose their savings, customers started a run on the bank. . . . Northern Rock's branches were poorly staffed and not used to a rush of cash withdrawals, and crowds built rapidly.
>
> TV images of worried savers queuing in the street were flashed around the world.[122]

As twenty-four-hour television news lingered on the queues that had formed outside branches of Northern Rock, bank shares fell and other lenders became embroiled in the crisis. Four days after the news broke, the UK government intervened by stating that it would guarantee all existing Northern Rock deposits. At this point, the queues finally disappeared.

Much in the unfolding of this event within the medium of television raises similar questions to those provoked by White's analysis of the *Challenger* catastrophe. When did the credit catastrophe actually occur? Was it when Robert Peston announced the bailout of Northern Rock, or when the US banks became too involved in subprime mortgages, or perhaps when Greenspan's policy really took hold (and when, exactly, was that)? The story of the credit crunch is far more historical than it is "eventful." It has to do with by now familiar bad banking practices and the ongoing deregulation of the global markets more than it does with any atypical occurrence—the financial crisis at Northern Rock, the investigation (televised live) of the bank's bosses by the Treasury Select Committee, the announcement that the Rock would be nationalized, the collapse of other banks and building societies in the United Kingdom, the United States (beginning with Lehman Brothers, Bear Stearns, Fannie Mae, and Freddie Mac), and throughout Europe. Although it has clearly been a real catastrophe for many businesses and individuals who have lost jobs and homes, it is inextricably tied to a banality of representation. There is a limit to how much a layperson can, or wishes to, learn about such things as securitization, debt bundling, junk bonds, liquidity, hedge funds, vanilla loans, quantitative easing, and so on.[123] If it is difficult to improvise the meaning of a rocket explosion, how much more vexing is it to vamp up the merger of two banking giants, HBOS and Lloyds TSB?[124] This is where Robert Peston (figure 2.1) comes in—precisely to give us *something to look at*.

Peston, it would seem, is oddly intriguing. More to the point, he has become something of a celebrity. There are biographies of him on YouTube, and even his detractors appear to be fascinated with him, or, more precisely, with the idiosyncrasies of his gestures and speech. He is, after all, *the face* of the credit crunch in the United Kingdom.

Figure 2.1
Robert Peston, BBC financial news.

Once the viewers were relieved of the less than riveting footage of people queuing in the street (figure 2.2), he became its point of location: its visual attraction.

Did Robert Peston Cause the Recession? Performing Economics

Let us return to the question we posed earlier: did Robert Peston cause the recession? Ways of trying to answer this question, and the debates they may open up, are of more interest to us here than any actual definitive answers themselves. If the ensuing debate turns out to be about causation, then the answer it receives—yes or no—will be about the functioning of a particular profession, journalism, and about whether people have faith in its claim to neutrality. A complex question will thus be rendered rather simplistic. In a *Daily Mail* article on Peston published a year after the crisis at Northern Rock, Michael Seamark places the BBC's claim that its financial reports are factual, responsible, and nonsensational alongside other claims that Peston is "a market menace" or, less sensationally, "the man who moves markets." The article is interesting not because it refers to Peston's second major scoop,[125] which "set off a fresh bout of chaos in the City," but because it refers to an earlier case of which Peston said: "I had thought I was merely doing an impartial reporter's job of describing Government thinking, based on conversations with ministers and officials in Downing

Figure 2.2
Northern Rock.
Photograph: Dominic Alves.

Street, the Foreign Office and the Treasury. But this was one of the rare occasions when a news story became a political event it its own right."[126] Clearly, then, Peston is aware of the possibility that a story may itself become an event.

For researcher of organizational behavior Mark Fenton-O'Creavy, the transformation of a story into an event is not something that happens on rare occasions but is rather a facet of social amplification and performativity. Media outlets, he suggests, can amplify risk, such as the risk of recession, by prioritizing their own demand for newspaper circulation or viewing figures. As we know, bad news sells. The selling amplifies the bad news. This argument of course assumes a representational model in which the fact of a recession exists outside of the mediation process, a discrete positioning that allows it to be affected—that is, to be subject to effects (in this case, to amplification). On the other hand, Fenton-O'Creavy also suggests that facts about the economy are themselves social, not given. They are therefore "true only so long as enough people believe in them." Can it really be the case that we believe ourselves into or out of a recession? As Fenton-O'Creavy writes in the Money and Management blog for the Open University: "What you believe does not just reflect our social world; it helps create it. *Performative* statements or beliefs are those which help bring about the conditions they describe."[127] Statements and beliefs about banks are therefore also seen as performative. If we say that banks are safe, "we help bring about the stability which makes this true." If banks believe that other banks are safe, they "ensure the stable operation of financial systems," which in turn helps to justify that belief. At the same time, if those beliefs change, and we withdraw trust from banks and banks withdraw trust from each other, "we help bring about conditions in which trust would be ill advised." Statements and beliefs about banks matter in two senses: they are meaningful and they have material consequences, placing an onus of responsibility on the media that broadcasters and journalists are often keen to deny, which they do by taking recourse to representationalism or the myth of disinterested detachment. But "whether they like it or not, journalists are not just reporting a financial crisis, they are performing it," according to Fenton-O'Creavy.

In *Do Economists Make Markets?* Donald Mackenzie and colleagues show how the notion of performativity breaks with the tenets of scientism, or "the widespread conception of science as an activity whose sole purpose is to observe and study, that is, to 'know' the world."[128] Economics, they argue "is not just about 'knowing' the world, accurately or not. It is also about producing it."[129] Their book offers a useful introduction to performativity in this context, outlining the contribution from Austin's original linguistic philosophy of performativity as well as from sociology, feminism, and science studies. Ian Hacking, for example, is positioned as having "showed how the science's representations of the world can be understood only in their close entanglement with intervention in that world."[130]

Mackenzie and colleagues are careful to distinguish between performativity and a simple, linear, deterministic model of cause and effect.[131] The question of causation is made complex within a notion of performativity informed by science studies and actor-network theory (ANT) in particular.[132] In ANT, the causal agent is regarded as being heterogeneous and multifaceted. The science and technology scholar and one of the key proponents of ANT, Michel Callon, applies this notion of causality to his argument about the performativity of economics. His thesis is that economics and the social sciences in general, like the natural and life sciences—dominated by describers as they are—actually "contribute toward enacting the realities that they describe."[133] The verb "contribute" is significant here because it qualifies and renders complex any simple model of causality.

For Callon, causation is not only heterogeneous but also nonlinear: "statements and their world are caught in a process of coevolution," he says.[134] Reality is thus "the temporary outcome of confrontations between different competing programs" or agencies.[135] Economies and markets in particular "are the temporary and fluctuating result of conflicts and the constantly changeable expression of power struggles."[136] Media, then, are part of that struggle for economic reality—at times confronting, at times collaborating with other agencies such as banks, regulatory authorities, and government. To paraphrase Bolter and Grusin, media have agency, but their agency is both constrained and hybridized. In other words they—and, for that matter, individual journalists—never act alone. Callon's reading of the performativity of identity suggests that if the identity of the journalist does not precede the performance or enactment of journalism, then that performance is corporeal as much as it is linguistic and incorporates a range of "sociotechnical" elements.[137]

To return to our earlier discussion of Robert Peston in the light of this analysis, we can certainly see that his performances are very clearly embodied—something that is made particularly explicit by what Brummer disparagingly refers to as Peston "oddly drawling," coupled with his "breathless" manner of speech.[138] Peston's media appearances mobilize a number of elements or agencies for their effect, but not solely through citation and the direct address to authorities such as the Financial Services Authority, the Bank of England, or the Treasury. We might just as well say that the policies and practices of these and other authorities—including the media and, specifically, the BBC—are what mobilizes Peston. They are the conditions of possibility for his performances as a journalist, performances that are therefore coperformances that coevolve over time but never reach their final point. Such an understanding of Peston's role and positioning in the constitution and performance of the media event that the credit crunch arguably was allows us to introduce and explore an alternative, more distributed and less centralized notion of agency, with agency defined primarily in relational rather than autonomous terms. This reframed understanding constitutes a challenge to humanism in questioning the assumption that any such event of the world is somehow all *about us*. The revised nonanthropocentric framework also requires a rethinking of the human's unique and driving role in bringing about these events. We may therefore want to turn again to Callon, who complicates Fenton-O'Creavy's understanding of performativity by distinguishing it from a human-centered rhetorical device of self-fulfilling prophecy. He writes, "Whereas the notion of a self-fulfilling prophecy explains success or failure in terms of beliefs only, that of performativity goes beyond human minds and deploys all the materialities comprising the sociotechnical *agencements* that constitute the world in which these agents are plunged: performativity leaves open the possibility of events that might refute, or even happen independently of, what humans believe or think."[139] In relation to the credit crunch, then, Peston might well have believed he was "merely doing an impartial reporter's job of describing," but he has, on this occasion, refused to realize that he was also producing, or rather coproducing, "a political event in its own right."

Recourse to Objectivity, or, "Don't Shoot Me, I'm Just the Messenger"

Does adopting the notion of performativity rid us of conceptual problems involved in representationalism? Although some theorists do think of it that way,[140] we—in a similar vein to Michel Callon and Andrew Pickering[141]—are not so certain about this. In as far as media events are performative, and have "nothing to do with balance, neutrality or objectivity,"[142] they do not preclude the possibility of a recourse to what remains a dominant claim within journalism and a dominant mode of understanding the world. For Barad, "the idea that beings exist as individuals with inherent attributes, anterior to their representation, is a metaphysical presupposition that underlies the belief in political, linguistic, and epistemological forms of representationalism."[143] The beings of which she speaks may be human or nonhuman, subjects or objects, people or things. The framework of representationalism allows us to think of their existence as something untouched by forms of representation. Representation, adds Barad, "is something explicitly theorized in terms of a tripartite arrangement."[144] This *spatial*, tripartite arrangement suggests that there are forms of representation (images and knowledge), objects of representation (things that are known), and representers (or knowers). Here, "it becomes clear that representations are presumed to serve a mediating function between independently existing entities. This taken-for-granted ontological gap generates questions of the accuracy of representations."[145]

Problematically for us, then, the questions that were generated in various media about the accuracy of the representations of the credit crunch treated the latter *as if it were an object* of representation that was entirely separate from the representational form of journalism and from journalists such as Robert Peston. As the target of many of these questions, Peston responded to them in a way that was resolutely defensive. He has defended himself as a journalist, he has defended journalism as a form of representation, and he has also, implicitly, defended the "metaphorical presupposition" Barad refers to—that all things, including the state of the economy, exist outside and independent of their representation. On YouTube; on his blog, Peston's Picks; and on the BBC News website and elsewhere Peston vociferously and repeatedly denies his involvement in helping to bring about the event he described by turning to the conventions of representationalism and away from a performative understanding of mediation. Over and again he cries, "Don't shoot me, I'm just the messenger." On his blog in particular Peston defends his own, as well as the media's, objectivity and object-like status, even while he condemns the banks—Northern Rock in particular—for irresponsible practices. From this point of view, the credit crunch had a cause for Robert Peston, but it had nothing to do with him or his profession. Yet media may be understood as the engaging but innocent middlemen personified and performed by Peston only within the conventions of representationalism. These conventions, as Barad suggests, center on spatialization: the creation of ontological gaps between "things." But as soon as time creeps in—and performativity, as Pickering argues, is a temporal concept that highlights material agential processes[146]—then claims to innocence as well as accusations of guilt are complicated in a way that seems to ensure that the show, which is always ultimately the describer's show, goes on.

To sum up our argument so far, we have posited that the credit crunch is a "media event with a difference," in that it highlights the role of media and mediation and *exposes the tension* between performative and representational accounts. We have also suggested that this tension is held in the otherwise rhetorical question of whether Robert Peston caused the credit crunch, and that it is relieved in his recourse to journalistic conventions, including the convention of representationalism. We have also proposed that the notion of performativity can offer a more productive and more dynamic understanding of what conventionally gets termed "media events" than the representationalist framework can. If Peston's announcement of the crisis at Northern Rock constitutes a case study in which the meaning and potentiality inherent in the concept of mediation are highlighted but ultimately closed down, then in our next case study—one that concerns the launch of the LHC project—mediation will play a central if not primary role. Yet, as we will show in the following section, this role and its transformative significance will become to a large extent obscured by the media focus on the nonrhetorical question of whether scientists will discover the God particle.

The Big Crunch: Mediating Life, the Universe, and Everything

LHC: The Aim of the Exercise

To smash protons moving at 99.999999% of the speed of light into each other and so recreate conditions a fraction of a second after the big bang. The LHC experiments try and work out what happened.[147]

The LHC project aims to recreate the Big Bang inside the world's largest supercollider. It is an experiment involving a—if not *the*—cosmic event that lies at the origin of life, the universe, and everything. The experiment is facilitated by a range of technologies (analog and digital) and by innovative programming and engineering that includes a 27-kilometer-long tunnel dug deep underground on the outskirts of Geneva. The goal is to collide counterrotating beams of protons, driven by large magnets and traveling at almost the speed of light, inside the tunnel and to then detect exactly how the universe came into being. Despite the trial runs, the "event" of a particle collision, with an energy of fourteen teraelectronvolts,[148] occurring at almost the speed of light, has not yet taken place. Although it is currently still wished for, it has already been and continues to be highly mediated. Our aim in this section of the chapter is to consider what mediation means in this context and then to analyze the relationship between the event and its mediation.

We want to propose that, in the context of the LHC project, the entangled relationship between events and their mediation is highlighted by virtue of being made extreme. Barad states that "to be entangled is not simply to be intertwined with another, as in the joining of separate entities, but to lack an independent, self-contained existence."[149] Events, in other words, are always already mediated. In this project specifically—one in which the full event has not yet occurred and will remain resolutely unrepresentable if and when it does occur—there is, more evidently, no event that is independent of its multimediation by agents ranging from magnets to computers, from particle physicists to the average web user.

Where the current status of a particle collision at CERN is something of a nonevent, this does not mean, in Baudrillard's sense, that it is condemned to being a pseudo-event: not only one that has not (yet) taken place, but one that will not take place at all. On the contrary, we will show that as a *mediated* rather than a *media* event, the collision is continually taking place. It is, in effect, virtually real. The concept of mediation cannot be equated with that of simulation precisely because the latter is predicated on an imploded division between the virtual and the real. If, for Baudrillard, the indivisibility of the event and its simulation in semiotic media signifies implosion, loss and devivification, for us the entanglement of events and their mediation signals a distributed, multiagential and embodied vitality that broadly conforms[150] to Bergson's notion of time, or life itself.

The question of the relation between the event and its mediation is paramount in this project. It contains within it a number of other relations that can be, or have been, regarded as oppositions, including those between representation and performativity; space and time; information and materiality; and liveness and liveliness. In order to sustain the question of relations and avoid both collapsing and opposing these phenomena, our approach here is based on division and reconciliation. Bergson's intuitive method, from which we freely borrow, states that we should: (1) identify the problem or distinguish between "true" and "false" problems; (2) distinguish between differences in degree and differences in kind; and (3) take time seriously.[151] The stated problem of the LHC project, the one foregrounded in its own publicity, concerns the presence (or absence) of a particle collision and the subsequent appearance of the elusive Higgs boson, or God particle. The Higgs boson is the quantum of a field that, if present, would explain why some particles acquire mass and others do not.[152] This is one of the so far unanswered questions in particle physics, and answering it would serve to reconcile the Standard Model of particle physics (effectively the Bible of particle physics) with Einstein's General Theory of Relativity.[153] The Higgs boson, if detected, would provide science with a theory of everything. It would link the physics of small and large objects, protons and planets. It would also afford ultimate, God-like knowledge, hence the veneration of this hypothetical entity. However, if such a particle actually exists, if it makes its reappearance in the fraction of a second following an automated event, then it will be so large and unstable that it will disappear before it can be visualized. All that could possibly be detected, therefore, is the "smoking gun"—*the animated lines or traces radiating from a point (i.e., the point of collision) on computer screens that, in representing the unrepresentable, are already contributing to bringing about the event that they subsequently describe.* The God particle will remain hypothetical, in the sense of being outside and above representational knowledge. Proof of its existence will be contingent, in large part, on our faith in the objectivity of visual media, or their ability to describe something from which they are epistemologically and ontologically separate. Presuming that this faith is at least combined with doubt in a secular, teletechnologized culture, the presence/absence of the collision-induced Higgs boson must give way to the ("true") problem of its mediation.[154]

Mediation is *a difference in kind* from media, to use a Bergsonian phrase, and one of the things that makes it so is its temporal nature. Mediation is primarily a temporal, multiagential phenomenon, a process rather than a spatialized and spatializing object. However, the division is not absolute, which is why we will attempt to reconcile as well as distinguish between the LHC as a media event characterized by spatiality and liveness on the one hand and the LHC as a mediated (non)event characterized by temporality and liveliness on the other. Time, for Bergson, is vitality, liveliness or life itself—but not liveness understood as the illusory movement of an object from A to B. We will show how *the illusion of movement is caught up with its actualization,* or how the proliferation of images is both performative of cosmic life and demonstrative of its elusiveness.

The quest for life, the universe, and everything at CERN seems to be more intuitive than intellectual, more concerned with affect than meaning.[155] It is embodied in part by anonymous particle physicists as well as by the celebrity TV presenter Professor Brian Cox (Peston's counterpart and the ever-smiling face of the LHC project). The language used to describe the project reveals connections between science and nature rather than signaling the conquest of nature by science. However, alongside CERN's "experimental metaphysics," to use Barad's term, and an affective investment in the project (and its failure, as we will show later in this chapter) to solve the mystery of the cosmos, it is possible to discern a far more instrumental and hard-headed rationale through which this particular science initiative has been framed.[156] The development of Grid technology (which is a form of distributed computing) at the LHC is a lesser-known aspect of the project. We will show how this less discussed aspect cuts across the dominant claim to experimentalism, open-endedness, and nonutility of the LHC project, thus serving as a salutary reminder that *mediation in this context is generative* but not necessarily progressive.

The Switch-On (and Off) of the LHC Project

The idea of a "media event" today, according to Merrin, "has permeated popular, academic and journalistic discourse, being variously used to describe the media's complicity in organized publicity, official events produced to be publicly broadcast, an undue prominence given to a minor news item or popular cultural phenomenon, stories involving major celebrity, or any story where the media's presence or the weight of coverage alone becomes noticeable."[157] The LHC project conforms to at least three of these definitions. The media were highly complicit in the organized publicity surrounding the launch on September 10, 2008. The moment when the supercollider was switched on was a preplanned, official event, produced to be publicly broadcast, and the weight of coverage was distinctly noticeable. We have already established that the LHC conforms to the conquest narrative outlined by Dayan and Katz and can even be regarded as the ultimate conquest (of nature). Even though it might also be described as a global media event, broadcast to the hundred countries that are reputed to have invested in the project, the BBC in particular[158] had a strong presence. Witness the discussion about the role of the media between writer and former particle physicist at CERN, Matthew Chalmers, and CERN's head of communications, James Gillies:

Rumour was that the switch-on date was arranged around BBC presenter Andrew Marr's holiday plans?

It's hilarious. The BBC did ask if we could put the date back if Andrew couldn't make it, and we said "no." But on the other hand, BBC Radio 4 pulled out all the stops and decided to do something unprecedented in science by devoting a day to the event, so in return we gave them a room just off the CERN Control Centre to use as a studio. The fact that Radio 4 went so big on CERN drove it out to the rest of the BBC, culminating in "Big Bang Day," and then out to the rest of the UK media and the world.[159]

The BBC's reward for publicizing the project was not just a room next to the control center but an event sufficiently live that it threatened or promised to fail—catastrophically. Chalmers's interview reveals that the switch-on had not been rehearsed and that the LHC's electrical circuits had not been tested. This lack of preparation resulted in an electrical fault that developed nine days later, causing extensive damage and necessitating the now notorious switch-off. Regardless of whether the incident was brought about by the pressure to meet a public relations deadline, the event's organizers subsequently struggled to retain control of the (negative) publicity, especially as this concerned the possible formation of black holes:

Did the black-hole Armageddon frenzy aid or hinder your communication efforts?

Ultimately it helped us by generating interest, but it also worried a lot of people and that makes me somewhat angry. People were phoning us up genuinely worried about the end of the world and demanding to know who CERN is accountable to.[160]

If the unofficial reaction to the public's irrational fears was, at times, rather tetchy ("Anyone who thinks the LHC will destroy the world is a twat"),[161] the official reaction, prior to switch-on, was to conduct a risk analysis. The LHC Safety Assessment Group (LSAG) was thereby able to report that LHC collisions "cannot be dangerous," and that microscopic black holes, should they form, would harbor "no associated risks."[162] Less proactive perhaps, and less playful than CERN's approach to Dan Brown's novel *Angels and Demons*[163]—"we put up a webpage and had fun with it"—the very existence of the LSAG (a reformation of the LHC Safety Study Group) testifies to the entanglement of the event and its mediation. Mediation, in this account, both incorporates and exceeds forms of media representation. The form most associated with the communication of actual or potential catastrophe is, as we have suggested, television. John Ellis, a scientist on the LHC project, is reported to have said in a blog by James Dacey: "Of the billions who tuned in for the switch-on, I suspect that many were only interested in seeing whether or not we would be blown to smithereens."[164]

White's argument for the reassessment of liveness in relation to catastrophe can also be applied to the LHC switch-on as a planned media event that promised catastrophe and delivered banality, along with the historical/futuristic and often spectacular details of particle physics. Vamping was much in evidence, mainly courtesy of the presenter Andrew Marr, who featured on BBC News 24 and throughout Radio 4's Big Bang Day. The challenge he faced is evident in the archived footage of the launch. YouTube features an eighteen-second clip from which Marr's vamping has been mercifully excluded:

"Five, four, three . . . "

As the countdown begins (we hear a male voice of European origin), the camera pans a room full of anxious faces, people biting their nails and staring at an array of computer screens.

"Two . . ."

The camera switches to the screens themselves. We see four screens, arranged in a block. The top left screen appears blank. Two others display graphs and simulations that are meaningless to the layperson. The bottom left screen resembles an early computer game; four T-bars face each other, waiting for a ball to bounce between them. This is where the action will take place. The anchor is laughing nervously at this point.

"Ha ha ha . . . "

"One"

"Zero"

"Nothing"

A small white ball has appeared in the centre of the screen, bottom left. It pulses and then fades.

"Yes!"

"Yes!"

We hear applause.[165]

Although there are no archives for BBC News 24, the BBC website carries transcripts and recordings of Big Bang Day, including the live coverage of the switch-on. At 8:30 a.m., John Humphreys, hosting the *Today* program, hands over to Marr at CERN[166]. Marr talks to CERN scientists about the background to the project while waiting for the countdown to begin. The protons, he says, "are champing at the starting gate." The project's success "is not a sure thing." While everybody in the control center "is staring at a series of screens above us," he asks how the launch will be effected: "What happens? Is it somebody hitting a return key?" Just then he realizes something has happened: "Oh, it's happened. The beams are into the collider." Is this an event? Apparently not. A stopper has been removed, but there will be a five-second delay in the acquisition of data: "There's five seconds in which nobody knows quite what's happening." He sees a vertical line. He learns that the beam has been stopped. He does not understand why. Then it is started again. "This," he says, "is a very, very hard thing to do." Protons will be behaving like kamikaze pilots, smashing into each other. "We're still waiting." Then the countdown begins, and Marr joins in. He knows "they've done it" only when he hears the applause.

The switch-on of the LHC project was therefore a media event of uncertain timing in which there was nothing to see. Nevertheless, it was undoubtedly a (tele)visual attraction consisting of both static images and replayed footage. There was, and still is, a great deal of space on show—terrestrial space of underground tunnels dug deep beneath the border between France and Switzerland and filled with machinery on an awesome scale, but also cosmic space of a past/future big bang. These attractions characterize news coverage of the switch-on,[167] television documentary,[168] and also the LHC website. The website offers an extensive photo gallery and an archive of the LHC's technological imaginary incorporating analog and digital machines, the industrial and the informatic, the spectacularly big detectors and the infinitely small objects of detection. Shown in figures 2.3 and 2.4 are the CMS detector and a mini-event (a collision, but not at full energy) that took place on December 14, 2009.

The photo gallery provides a link to an animation that is frequently replayed in other visual media (see figures 2.5 and 2.6).

It begins with a graphic representation of the circular 27-kilometer collider and the slingshot structure that will accelerate proton beams in opposite directions until they reach virtually the speed of light and collide inside one of the four detectors—in this case, ATLAS. Upon pressing "Play," a single red proton enters the circuit, shortly followed by another. They pursue each other around the miniature rings before being propelled in opposite directions around the main loop of the collider. Each proton makes several high speed laps before they both enter the detector, where we zoom in to watch them square up to each other, Western-style, not kamikaze—this is one big showdown. They advance, they collide, they *transform* into a brilliant display of radiating lines and colors that burst out in all directions and appear to break through, to *transcend* the boundaries of the ATLAS detector.

Enchanted Objects

Here then is the smoking gun, the only evidence there will ever be of the elusive Higgs boson, the so-called God particle created by the collision between two proton "bullets." Even though the event has not yet happened, the ATLAS detector has already performed and, in numerous animations, continually performs its task of detecting the undetectable so that TV screens and computers everywhere—in the control center and in homes—can represent the unrepresentable. For Baudrillard, the playing out or precession of a scenario before it can occur empties the event of meaning and negates it.[169] However, unlike in the case of a war which follows a war film, or a nuclear explosion that comes after a film about a nuclear explosion, *there is no event of a*

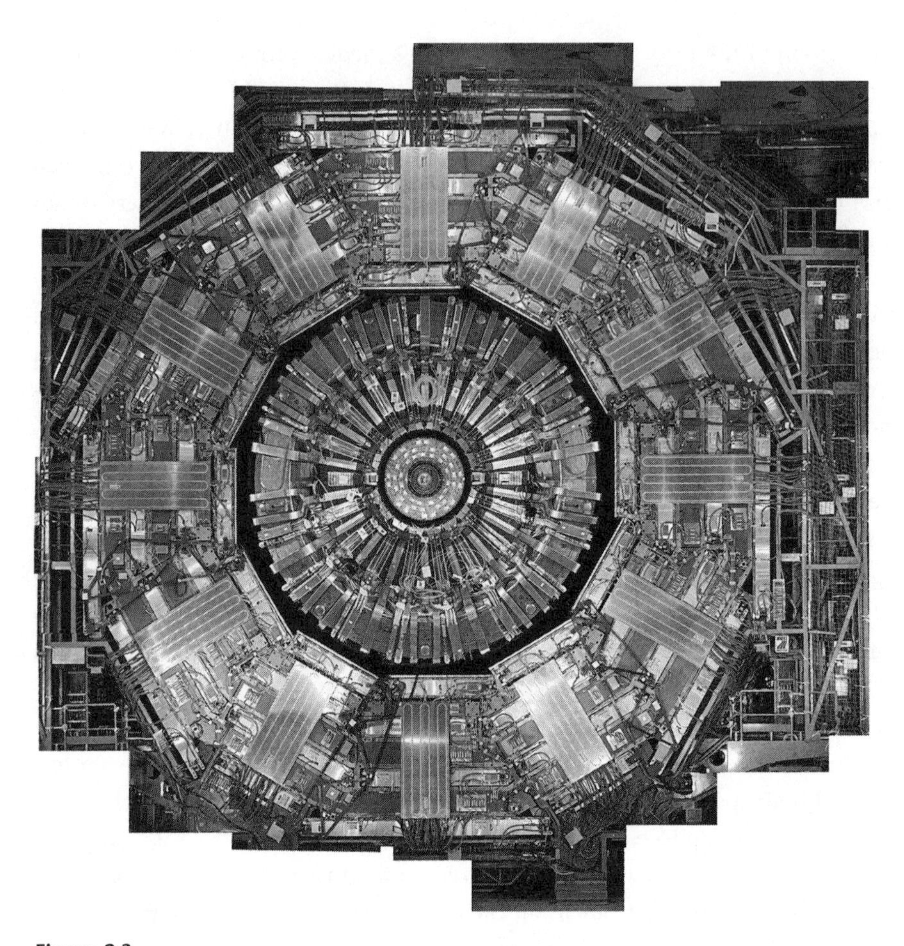

Figure 2.3
CMS giga panorama landscape.
Photograph: Maximilien Brice, Maurice Amoos, Michael Hoch, Maaf Alidra.

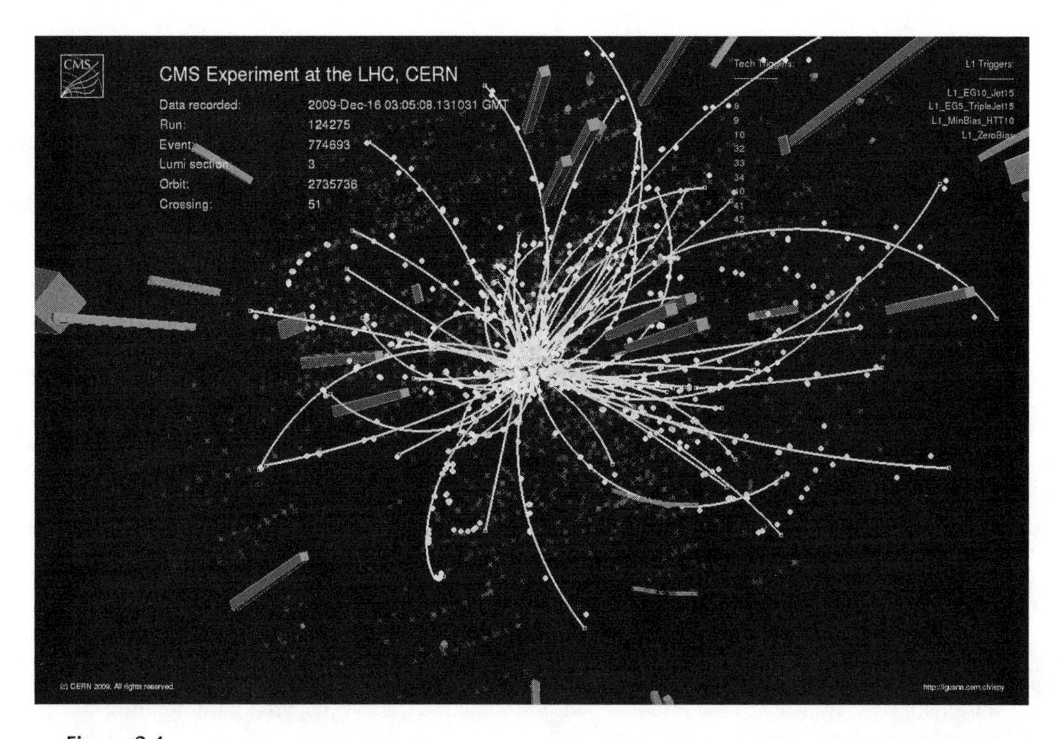

Figure 2.4
A mini-event or particle collision detected by the CMS on December 14, 2009; LHC Machine Outreach.

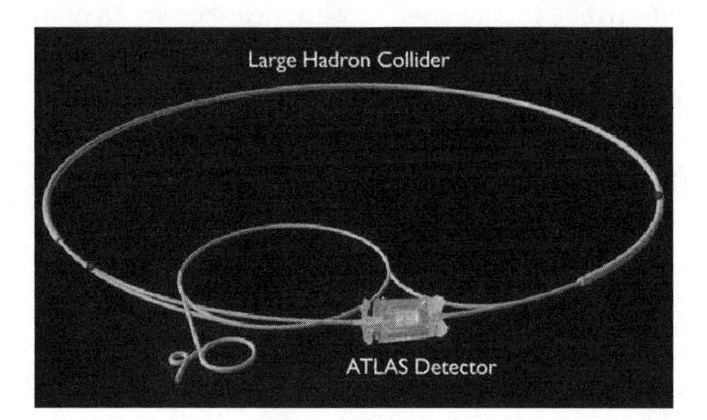

Figure 2.5
Animation showing particle collision in the ATLAS detector, LHC Machine Outreach.

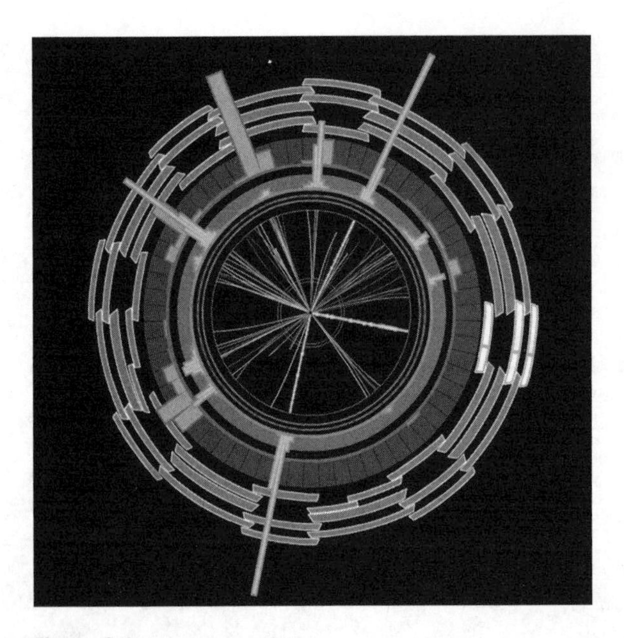

Figure 2.6
Still image from the animation showing the "smoking gun."

collision-induced Higgs boson (catastrophic or otherwise) that is discernable in space and time from its animations. Are these animations, then, no more than a "disastrous interference" in history, brought about by the over-proximity or implosion of the event and its mediation)?[170] Are they, effectively, the deterrents of the real—or are they in some sense *it*?

We would argue that in a representational sense, they are neither. Representational realism, or representationalism, "is the belief in the ontological distinction between representations and that which they purport to represent; in particular, that which is represented is held to be independent of all practices of representing."[171] Representationalism presents us with autonomous entities such as particles that, by virtue of being independent of our scientific and other ways of seeing and knowing, can be represented either more or less accurately. The LHC project is over-invested in scientific representationalism and therefore primarily perceives the event as an object, or an autonomous thing, rather than a phenomenon or process incorporating multiple agencies. The event as an object or thing has a beginning (a bang) and an end (a universe), and it can be represented in science and the media because it preceded the intervention of science and the media. Ultimately, the event becomes, in the word of the anthropologist Alfred Gell, "enchanted";[172] it is elevated, as Lucy Suchman puts it, to the status of "something that *is* and has effects, rather than as an effect in itself."[173] This enchantment can be manifest in verbal as well as visual images. It need not apply specifically to the collision event itself but also applies to anything or anybody that can stand in for it: the project, a person, a proton.

The LHC website includes a list of enchanting facts, ranging from the sublime to the ridiculous:

> The combined strands of the superconducting cable being produced for the LHC would go around the equator 6.8 times. If you added all the filaments of the strands together they would stretch to the sun and back 5 times with enough left over for a few trips to the moon.
>
> Part of the LHC will be the world's largest fridge. It could hold 150,000 fridges full of sausages at a temperature colder than deep outer space.[174]

Protons, as we have seen, are particularly daring and heroic and, as such, they are anthropomorphically interchangeable with people:

> When protons arrive in the LHC they are travelling at 0.999997828 times the speed of light. Each proton goes around the 27 km ring over 11000 times a second.
>
> A nominal proton beam in the LHC will have an energy equivalent to a person in a Subaru driving at 1700 kph.[175]

In place of representationalism and its enchanted objects, Barad "turns" to a notion of performativity that might apply to the natural as much as to the social world. She writes, "Performative approaches call into question representationalisms' claim that there are representations, on the one hand, and ontologically separate entities awaiting representation, on the other, and focus inquiry on the practices or performances of representing as well as the productive effects of those practices and the conditions for their efficacy."[176] Without wishing to validate this or indeed other attempts to identify a conceptual turn, we might, by acknowledging Barad's place in a legacy of debates on performativity and against representation, think with her about the productive effects of the LHC's representing practices. To what extent are they helping to bring about the event that they subsequently describe?[177] A performative account of science (and, indeed the media) "takes account of the fact that knowing does not come from standing at a distance and representing but rather from a direct material engagement with the world."[178] What is the nature of these direct material engagements? Particles are produced experimentally and visually, as entities and as images,[179] by what Barad calls "a Cartesian habit of mind"—one that atomizes, isolates and divides things along a subject/object axis. Bergson similarly criticizes the tendency of what he calls "intellectual knowledge," which, by cutting and isolating objects from among processes, converts movement to stasis, time to space. From this perspective, a particle is not something that *is* and *has effects*, but rather *an effect in itself*—an effect of a habit of mind. For Bergson, we can change this habit of mind by thinking more intuitively, less intellectually, more in terms of our connection with than our separation from the world around us. As Elizabeth Grosz explains, "Bergson claims that there are two ways to know: one, intellectual, immobilizes and isolates, facilitates practical action and use but thereby moves from the real to its schematization; the other, intuitive, seeks continuity, indiscernability, flow, and duration, immobilizes practical action but brings us directly into connection with the dynamism of the real."[180] The productive effect of the LHC's representing practices is both to repeat the Cartesian intellectual habit of mind and to challenge it by taking us back to the beginning of time. If the Large Hadron Collider itself is a time machine for enchanted objects, then their animation plays out the possibility that they could be transformed into something that transcends or remains elusive to detection, to current and future forms of representational knowledge.

Mediated (Non)Events, Time, and Lifeness

We have argued that mediation is not synonymous with simulation and that it exceeds representation to encompass the performative relationship, the "intra-active" relationality or, as Pickering would have it, the "mangle" between entities held separate by our dominant habits of mind. Mediation is not a spatialized thing or object but first of all a temporal phenomenon, a process involving human and machinic agencies. Like all processes, it cannot be represented, though it can be apprehended intuitively. Mediation is key to the vitality of the LHC nonevent that has not actually taken place but is virtually real—material, agential—and continually taking place. Our reading of this nonevent entails retracing Baudrillard's steps back to Bergson: "Whereas the virtual was once that which would become actual, now, for Baudrillard, it is that which deters it."[181] In signaling that the relation between the event and its mediation is paramount in the LHC project, and in identifying the project as a nonevent, we are placing our argument firmly within a Baudrillardian frame. According to Merrin, Baudrillard's discussion of the nonevent is significant because it "reveals a mediatic operation extending beyond the mere production or inflation of 'news,' one encompassing and implicating our entire epistemological relationship to the wider world."[182] For us, that epistemological relationship is also an ontological one. Yet in Baudrillard's account it remains, even in the context of the implosion of binaries, too Cartesian, too intellectual—and, ultimately, too nihilistic.

What is at stake in proposing, instead, a performative and productive entanglement between media and events is precisely their shared vitality. If a programmed event cannot constitute the event's occurrence, it can co-constitute it. In this way, *the animation of a particle collision—itself a manifestation of liveness, an illusion of movement, an instance of representationalism—becomes a vital part of its reality, and instead of constituting an interference in history, it contributes to the constitution of its metamorphosis in and through technological forms*. The stage for the in/action of events has been shifted and expanded, but not removed. Events are not imploded in media; instead, they are transformed through mediation. Mediation does not posit a reciprocal end game—the "dissolution of TV into life, the dissolution of life into TV"[183]—but rather a productive relationality that is not an end point in history, but part of its creative evolution.

Like the H1N1 flu pandemic, the Iranian postelection protests or the atrocities at Abu Ghraib—all "superconductive events," in Baudrillard's terms[184]—the LHC events are multimediated and, to use Bolter and Grusin's notion, hypermediated. Hypermediacy is in part an aesthetic, a collage effect of different media forms and styles that often come together on a single screen.[185] But the concept of hypermediacy also captures McLuhan's assertion that the content of media is always other media. Mediation, for Bolter and Grusin, is always the remediation of old media in other, newer media. More than that, (re)mediation is intrinsically connected to wider social and economic forces, so the question of agency is never simple, never wholly human or technological, but rather hybrid, distributed (if not evenly) and processual. Another way of putting this would be to say that as a hypermediated event, the LHC is not, and never will be, fully automated or wholly machinic. It is composed, as we have suggested, of machinery big and small, analog and digital, but also of massive financial investment as well as personal and professional investment measured in career lifetimes. Of all the particle physicists, amateur and professional, who embody the nail-biting aspirations of the LHC project, one has been chosen to publicize it, to bring the God particle to the people.

Things Can Only Get Better?

As the keyboard player for D:REAM, Brian Cox was one of the musicians behind New Labour's 1997 election anthem, "Things Can Only Get Better." His faith in progress has been rewarded with enhanced celebrity status as the face of the LHC project and other scientific wonders.[186] In the absence of an actually unfolding televisual event,

Cox, ex–pop star, gives us something to look at. As he races around the 27 km circuit, helmeted, adventurous, gleeful, he becomes interchangeable with the protons that enchant him, the sixteen particles of the Standard Model whose names—muons, gluons, and so on—he lovingly recites. He affectively narrates the impasse in physics, and if he is excited by the prospect of detecting the Higgs boson, he seems even more excited by the possibility that it will not be detected and therefore does not perhaps actually exist. "This would provoke a revolution in physics," he proclaims.[187]

Cox, among other scientists at CERN, stands more than ready to break with his habits of mind. If the God particle would join together their seemingly intractable scientific models and complete their intellectual knowledge of life, the universe, and everything, then long may it continue to elude them. This seems likely, the BBC Four documentary suggests, as physicists are not entirely sure what a particle is—a point in space or an extension in time. If particles were more like strings: "not the idealized points of Euclidean geometry, but . . . objects extended in one . . . or . . . more dimensions," then they would be more amenable to gravity, the force that underlines Einstein's General Theory of Relativity, but is at odds with the Standard Model of particle physics.[188] String theory, like the Higgs boson, *could* bridge the two models, except that strings are so "unimaginably small" that it would be impossible to build a collider big enough to detect them. The existence of superstrings is beyond, if not above, representation even by means of the smoking gun. String theory is therefore more metaphysics than physics; it is an intuitive rather than intellectual philosophy of life that is characterized by a desire for connection, not conquest. Its presence in the discourse of the LHC project helps to reinforce an aura of creativity, experimentalism, and nonutility—a sense that the whole project has been designed, funded, and executed for the sake of pure knowledge.

The claim to pure knowledge is undermined in this context by the LHC's subproject: the development of the Worldwide LHC Computing Grid (WLCG). The WLCG is defined as "a novel globally distributed model for data storage and analysis."[189] CERN needs to distribute and decentralize its computing because the LHC project will generate a phenomenal volume of data. So, just as "the Web was invented at CERN so that scientists could share information,"[190] the Grid will be developed so that information can be stored and analyzed in different locations. If the growth and popularity of the Web was unforeseen, the potential applications of Grid technology are not. The idea is that Grid technology will turn computing into a utility like gas or electricity—a utility that is, at the same time, infrastructural.[191] This result will be brought about through the development of middleware, a layer in between operating systems software (like Windows) and applications software (like Microsoft Word). This layer will enable Internet access to be regulated through automated systems of tracking and security, expressly designed to prevent the free flow of information, file sharing, and other activities attributed to "malicious hackers."[192] The mission, currently, is to standardize computing grids so that business, government as well as scientific and academic investors can "influence and outreach to end-users" and "gain visibility for brand, products" and so on.[193]

Grid computing is the relatively invisible prospect of the LHC project, and although it is continuous with the ongoing marketization of the Internet that literally puts pay to its early utopian promise—including the promise that it might constitute a virtual life-form, that is, an autonomous, information-processing, self-organizing, nonorganic system—it is discontinuous with the LHC's intuitive philosophy of life.

Sensing Media Metamorphosis

To sum up, in this chapter we have drawn primarily on an alliterative trilogy of thinkers—Baudrillard, Bergson, and Barad—in order to analyze the LHC project as a contemporary exemplar of the problem of mediation. "Mediation" is a term that has begun to be interrogated in the context of new media studies but has yet to be fully explored, at least in its temporal aspects. Too easily reduced to a synonym for media representation, and therefore aligned with representationalism, mediation is understood by us instead as a difference in kind—performative, material, vital. It is neither reducible to, nor separable from, relatively stabilized and spatialized media forms in which vitality has been said to be illusory or eliminated.

Although we have described the LHC project as a nonevent—because it has not happened yet and because it is unrepresentable—we have also challenged Baudrillard's reading of the nonevent as the outcome of the implosion of the symbolic in the semiotic, the real in the virtual. *By replacing the metaphor of implosion with that of entanglement, by regarding events and their mediation as co-constitutive, we have also, effectively, reinstated becoming rather than deterrence as a key aspect of mediation.* If we have sought to recognize Baudrillard's contribution to debates on media events, we have also incorporated Bergson into our critique of Baudrillard in order to argue that mediation is a vital process, one that produces rather than merely constructs the real.[194] Just as a collision event at CERN is not likely to produce black holes and a cosmic catastrophe, the animation of a collision event does not effectuate its (for Baudrillard, catastrophic) negation. The relation between the real and the virtual in this context is, we have suggested, both indistinguishable[195]—in that there is no event of a collision-induced God particle that can be distinguished from its representation—and cooperative. Despite an overinvestment in the presence/absence of the Higgs boson and other enchanted objects, what CERN makes operational is the becoming of life, the universe and everything *in*, or rather, *as* processes of mediation.

Mediation incorporates technologies and their users, machines and their human counterparts. It is multiagential, and rather than constituting an interference in history, we have argued that it is contributing to its metamorphosis. Although we do not subscribe to Durkheim's "absolute" distinction between the sacred and profane as a religious method of social structuring, or to the ways in which this has been applied to the media (media is not a religion!), we would not want to characterize mediation as an unequivocal good, as becoming without limits, a process without stasis or a potentiality without closure.[196] Its vitality is perhaps only the flip side to what Baudrillard refers to as "fascinating, indiscriminate" virulence.[197] Though mediation cannot be anthropomorphized and instrumentalized (in the same way that technology cannot, according to Heidegger), the media to which it is tied continue to be diverted to political and economic ends. The LHC project promotes as well as experiments with physics, it works to further commodify computing even as this stretches back to infinity—and beyond.

The credit crunch and the "big crunch" of the LHC project have been positioned by us here as two generically different media events that are linked by virtue of being actually or virtually catastrophic, and at the same time, quite ordinary, even banal. In both cases, we have sought to differentiate but also reconcile the role of media and mediation, spatiality and temporality, liveness and liveliness. In doing this, we have been guided by Bergson's intuitive method that enables us to understand them as related but nevertheless irreducible phenomena—precisely as differences in kind. If media such as television and the press have been associated primarily with representationalism, then we have attempted to open out the question of mediation by associating it with performativity. However, we have attempted to avoid creating a "false" division between representation and performativity, just as we have tried to avoid conflating mediation and performativity. We are talking here about alliances and associations, not equivalences. Mediation is not only performative and temporal but also incorporates spatialization and representationalism—an effect of our Cartesian and intellectual habits of mind and a dominant form of knowledge.

If the credit crunch and the big crunch both center on events that are entangled with their mediation, one of the differences between them is that of degree. We have argued that the relation between the event and its mediation is *a* central question in the context of the credit crunch, but *the* central question or problem at the heart of the LHC project. Where the former enables us to think about the *performativity of the event*, the latter, we might say, obliges us to think about the *event as performativity*. Where there is no distinction (in space or time) between the event and its mediation in the LHC project, we have nevertheless refused Baudrillard's invitation to think in terms of implosion, devivification, and a disastrous interference in history brought about by the ubiquity of media. Media, in our account, do not constitute a black hole that obliterates events. They do not swallow up catastrophes any more than they create sacred ceremonies that have a unifying effect on the world out there. *Media, through processes of mediation that incorporate and exceed them, contribute to the generation of the world out there*. Their role is transformative as much as it is informative, but *this metamorphic role is first of all something we sense rather than something we see*. Understanding it therefore requires less intellect and more intuition:[198] it requires a willingness to suspend our habits of mind, to garner insights that might then be subject to further critique.

3 Cut! The Imperative of Photographic Mediation

"What Is Photography?" Yet Again

As the event of mediation is, like time (or, indeed, life itself), both invisible and indivisible, any attempt at its representation must ultimately fail. In this chapter, we offer a challenge to representationalism by looking, perhaps somewhat counterintuitively, at a form of media practice that is most readily associated with representationalist ambitions: photography. Our aim is not so much to raise familiar questions regarding photography's truth claims and its supposed "indexicality," that is, the relation the photographic image allegedly maintains to an object it is said to represent. Rather, we are interested in foregrounding the productive and performative aspect of photographic acts and practices that are intrinsic to the taking or making of a picture. With a view to this, we propose to understand photography as an active practice of cutting through the flow of mediation, where the cut operates on a number of levels: perceptive, material, technical, and conceptual. The recurrent moment of the cut—one we encounter not just in photography but also in film making, sculpture, writing, or, indeed, any other technical practice that involves transforming matter—will be posited by us as both a technique (an ontological entity encapsulating something that *is*, or something that *is taking place*) and an ethical imperative (as expressed by the command: "Cut!"). The key question that organizes our argument is therefore as follows: if we must inevitably cut, and if the cut functions as an intrinsic component of any creative, artistic, and especially[1] photographic practice—although this is still only a hypothesis—then what does it mean to *cut well*?

This study of the cut as an inextricable accompaniment of mediation will be enacted as an encounter between the two philosophical traditions that are shaping the argument of this book: the vitalism of Henri Bergson and Gilles Deleuze on the one hand, and the *différance* of Jacques Derrida and Bernard Stiegler on the other. Yet, as explained earlier, this philosophical encounter will be of interest to us only in so far as it will allow us to move the debate on media and mediation on, not as an exercise in philosophical point scoring. Contrary to Bergson and Deleuze, we will see photography as more than a series of frozen "snapshots." *In introducing a distinction between photography as a practice of the cut and photographs as products of this process of cutting, we will also aim to capture and convey the vitality of photographic movements and acts*. If, indeed, "To live is to be photographed,"[2] then, *contrary to its more typical association with the passage of time and death, photography can be understood more productively in terms of vitality, as a process of differentiation and life-making*. It is, paradoxically, precisely in its efforts to arrest duration, to capture or still the flow of life—beyond singular photographs' success or failure at representing *this* or *that* referent—that photography's vital forces are activated, we will claim.

To see this process of cutting in action, in the final part of this chapter we will look at practices that use "the cut" as both a subject and a method. Focusing on the photographic series *Oblique* by the Australian artist Nina Sellars, we will raise some broader questions about visualizations of the open and wounded body in the current media culture. Specifically, we will be interested in interconnections and symbolic transactions between widely distributed media images of open bodies in franchised TV makeover shows such as ABC's *Extreme Makeover* and MTV's *I Want a Famous Face*, and gallery-destined photographs of what we can call "mediated pain." Through this, we will aim to identify points of convergence between art, philosophy, and science, as well as trace some of the ways in which creative practice can alter reality by intervening in it on a material level. The way "the cut" is variously exercised by the artist, the surgeon, and the philosopher will become a focal point for a discussion about photographic mediations.

Life Will Find a Way

New York fine art photographer Joel Sternfeld's images from his *Walking the High Line* series (2000–2001)[3] depict what looks like an abandoned railway track amid an urban landscape, with thick vegetation of all sorts—grasses, shrubs, trees, wildflowers—enveloping the metal and wood of the line. The color palette of the photographs changes depending on the season when they were taken: from verdant green through to autumnal brown and snowy white. The images could be said to encapsulate the oft-cited line by Dr. John Hammond, the owner of Jurassic Park, in the sequel to the eponymous movie: "Life will find a way." The High Line itself is a multilayered space. Opened in 1934, this elevated railway line had been designed for the delivery of milk, meat, and other raw and manufactured goods to the stores of Manhattan, a purpose it served until 1980. In an interesting remediation of the city's history, geography, and architecture, it has now become an urban park, providing a unique vista of New York City from the height of thirty feet while also staging an ongoing encounter between the living and the technological. This encounter is itself mediated by a series of image- and sound-based artworks, with Richard Galpin's "Viewing Station" offering a particularly striking take on the space (figure 3.1). Consisting of a viewing device and a large white rectangular board containing a number of cutouts (figure 3.2), it encourages walkers on the High Line to see this particular corner of Manhattan otherwise, through a series of incisions and cuts that transform the familiar view into a modernist abstraction.

It is particularly apposite that a specially commissioned piece by Galpin should make its way into the High Line, given the artist's long-standing engagement with

Figure 3.1
Richard Galpin, "Viewing Station" (2010).
Photograph by Joanna Zylinska.

Figure 3.2
Richard Galpin, "Viewing Station" (2010)
Photograph by Joanna Zylinska.

reimagining city spaces through a series of interventions and cuts, going back all the way to his "peeled photographs" from 2001 ("Esso," "Beyond This Point," "Signs, Blinds, Fabrications"), but also evident in his more recent "Splinter" works (2009–2010). Significantly, they are cuts made "at one remove," so to speak, as Galpin slices with a scalpel into photographic representations of urban spaces, peeling away "the emulsion from the surface of the photograph to produce a radical revision of the urban form."[4] In "Viewing Station," he alters his established process, with the High Line itself—both as a location and a familiar photographic representation of it—becoming a space of incision from which to see things in a different light. In this way, the High Line also becomes a site of what in this chapter we want to term "photographic mediation," whereby an act of making cuts is taken on, and foregrounded in the photographic process, for aesthetic as well as ethical reasons.

Through the Chaos

In the concluding pages of *What Is Philosophy?* Deleuze and Guattari provide an overview of their theory of concept-formation as a way of organizing the turmoil of the world, of taming its vibrations. (Let us note in passing that these "vibrations" are not just metaphorical: physics tells us that particles of matter are in constant movement at both classical and quantum scales, although what it actually, practically, means on a day-to-day basis is of less concern to us at the moment.) Deleuze and Guattari write, "Chaos has three daughters, depending on the plane that cuts through it: these are the *Chaoids*—art, science, and philosophy—as forms of thought or creation. We call *Chaoids* the realities produced on the planes that cut through the chaos in different ways."[5] To explain how we deal with the physical reality of the world, Deleuze and Guattari conjure up these three mythical creatures that stand for different ways of organizing matter, physically and conceptually, into forms.

Following their lead, we want to suggest that the practice of cutting is crucial not just to our being in and relating to the world, but also to our *becoming-with-the-world*, as well as *becoming-different-from-the-world*. It therefore has an ontological significance: it is a way of shaping the universe, and of shaping ourselves in it. Indeed, Deleuze and Guattari themselves claim that a concept, any concept—which for them does not serve "to replicate accurately in discourse specific segments of the world as it really is (as science does), but to propose articulations of and/or solutions to problems, to offer new and different perspectives on orientations toward the world"[6] is a "matter of articulation, of cutting and cross-cutting."[7] The process of cutting is one of the most fundamental and originary processes through which we emerge as "selves" as we engage with matter and attempt to give it (and ourselves) form. Cutting reality into small pieces—with our eyes, our bodily and cognitive apparatus, our language, our memory, and our technologies—we enact separation and relationality as the two dominant aspects of material locatedness in time. Now, whether these enactments get recognized as legitimate daughters of chaos such as art, science, and philosophy (at least in Deleuze and Guattari's kinship structure) or whether they remain nameless, unlawful, half-bred, depends on the mobilization of the whole network of institutional responses that give names to cultural practices. What is interesting for us here is not so much the particular acts of legitimate recognition but rather the very process of making cuts—both on the level of matter (the work of Chaoids and their bastard siblings) and on the level of culture (the work of institutions that give names to things and hence authorize their existence).

Photography fits perfectly into this dualist schema, due to its uncertain ontological status: it is both an act of carving the world into manageable, temporarily stabilized two-dimensional images of it and a set of institutions and conventions that arbitrate over doing things with a camera in a variety of different contexts. However, photography is more than the sum of solidifying acts and stabilized objects. Mobilizing the thinking on time, movement, and creation by two philosophers who did not have many positive things to say about photography—Bergson only ever talks about "snapshots," which stand for him for compromised attempts to give form to duration, and Deleuze is clearly much more interested in cinema—we want to look more closely at this elision of photography from the materialist philosophy of duration in search of a more productive engagement with "the photographic cut." We also want to explore the inherently creative potential of photographic practice, as something that exceeds the realm of human creativity. Hollis Frampton articulates this potentiality of photography, realized in the act of making cuts, in the following terms:

> The photographer's whole art may be seen as a cutting process. The frame is a fourfold cut in projective space. The selection of a contrast curve of a given slope and shape, and the mapping of bright and dark zones on that curve, are clearly operations that cut the intended from the possible. If I am making a color print and, at a certain point, decide that it must be lighter, more green, higher in contrast, then I am making a threefold cut in an unmodified field of fifty-five possibilities.[8]

There is always more to photography than a series of photographs already in place: its potentiality includes all the photographs "not yet taken," all those cuts "not yet made." In making incisions into duration, in stabilizing its flow into graspable entities, photography is inherently involved with time.

Most people's everyday experience positions them as collectors of memories, viewers of moments of captured temporality, and producers of such moments. Arguably, over the last half century, photography has become so ubiquitous that our sense of being is intrinsically connected with being photographed, and with making sense of the world around us through seeing it imaged. As David Bate points out, "If I wish to travel, a photograph of my face is required to indicate my identity in my passport, without which it would be hard to go anywhere. I am likely to have already seen photographic images of my destination before I have even been there."[9] Yet even though photographs are indeed ubiquitous and even if their primary mode of functioning is that of recording (the passage of) time, and of introducing a differentiation between the now-time and the-time-that-once-was, this relational, dynamic aspect of photography arguably gets lost among the plethora of photographic objects and artifacts. This problem is exemplified in Barthes's position as presented in *Camera Lucida*, whereby his search for the essence of his dead mother in the Winter Garden photograph, which supposedly represents her (and which we never see), "supersedes, and overlays, his search for the essence of photography. It leads him to replace photography with the photograph, memory with *the* memory, virtual existence with actual existence, and, ironically, perhaps, (her) life with (his) death."[10]

Intuition and Time

However, if we are to think about photography in terms of mediation—whereby mediation stands for the differentiation of, as well as connection between, media and, more broadly, for the acts and processes of producing and temporarily stabilizing the world into media, agents, relations, and networks—we need to see the ontology of photography as predominantly that of becoming. The best way to understand becoming is arguably through a Bergson-inspired concept of intuition. Becoming as an inherent condition of photography must be differentiated from narratives about photography's alleged transformation, evolution or even death in the digital age, that is, about photography "becoming different from what it once was." Narratives of the latter kind are premised on separating one aspect of a phenomenon—a technical one—and then extrapolating its role to that of a determining factor in the development of a medium. Yet, as argued in chapter 1, it is both false and reductive to propose such a cause-and-effect model of understanding digital media, whereby technology is treated as a mere instrument, a discrete entity that can be said to have one-way "effects" on other entities, rather than as an environment or a field of forces. This latter understanding of technology from the midst of its force field, as it were, can first of all be intuited rather than grasped cognitively. After Bergson, intuition stands for more than mere feeling or inspiration: it is a form of knowing through practice. The concept of intuition is close that of instinct, which can be defined as "the unconscious bond with reality that provokes bodily action and reaction in all animals."[11] In humans, instinct is supposed to have subsided just as intelligence has developed. For Bergson, intuition offers a means by which to return each to the other.

Bergson's turn to intuition is an attempt to recapture this not quite yet atrophied instinctive mode of relating to reality that does not posit a prior separation between the knower and its objects in the way the intellect does, but which rather presents knowledge to us "from within." For Bergson, "the intellect aims, first of all, at constructing. This fabrication is exercised exclusively on inert matter, in the sense that even if it makes use of organized material, it treats it as inert, without troubling about the life which animated it. And of inert matter itself, fabrication deals only with the solid; the rest escapes by its very fluidity."[12] The problem with the intellect is that it "cannot, without reversing its natural direction and twisting about on itself, think true continuity, real mobility, reciprocal penetration—in a word, that creative evolution which is life."[13] It is therefore incapable of recognizing any real novelty, or what Bergson calls "absolute becoming." Indeed, the intellect is "awkward the moment it touches the living."[14] It can only grasp what it already knows; it only knows "the repeated." Although the intellect has a "natural inability to comprehend life,"[15] instinct and intuition (whereby intuition for Bergson is a form of conscious or refined instinct) are "molded on the very form of life."[16]

The Ghost in the Cinematograph

The intellect's manner of functioning is mechanistic: it "represents becoming as a series of states, each of which is homogeneous with itself and consequently does not change."[17] In other words, the intellect cuts up reality into fragments, which it then passes off as truthful representations of this reality. Bergson explains this working of the intellect in our attempt to make sense of the world as follows:

> Instead of attaching ourselves to the inner becoming of things, we place ourselves outside them in order to recompose their becoming artificially. We take snapshots, as it were, of the passing reality, and, as these are characteristic of the reality, we have only to string them on a becoming, abstract, uniform and invisible, situated at the back of the apparatus of knowledge, in order to imitate what there is that is characteristic in this becoming itself. Perception, intellection, language so proceed in general. Whether we would think becoming, or express it, or even perceive it, we hardly do anything else than set going a kind of cinematograph inside us.
>
> We may therefore sum up what we have been saying in the conclusion that the mechanism of our ordinary knowledge is of a cinematographical kind.[18]

For Bergson, the cinematograph as a machine for animating still images becomes a metaphor for the wrong or false kind of perception, as its mode of working is premised on the forgetting or overlooking of duration. (The cinematograph will later be redeemed by Deleuze as offering precisely an insight into duration, although Deleuze will remain in agreement with Bergson about the latter's critique of the elision of becoming from the process of looking, a process which is focused only on "states." Photography for Deleuze is just a form of "molding"; it is a way of conserving immobility, and the cinematic image is a "mobile section" or a "modulation."[19])

It may seem from this argument that our intellect can understand photographs but cannot understand photography; it can understand media but not mediation. However, Alexander Sekatskiy makes an interesting observation about what it is that Bergson actually sees as passing through the cinematograph of our mind, and about the ensuing misreadings of this passage in many commentaries on photography and cinema. He suggests that the "photographs" that Bergson is talking about "are not photographs: they are much more like film-stills. By mixing up the incommensurable temporalities of photograph and film-still we fail to distinguish between images of perception and images of memory. Within the duration of cinematic time a still photograph is a completely lifeless, alien object, and by the same token, a film-still is dead when it is placed among photographs."[20]

Before we raise some further questions about such a (misguided, it seems) reduction of photography to a mere illustration of what is wrong with our perception of the world and propose to explore photography's own intrinsic movement and lifeness, we would like to take a closer look at this recurrent use of the notion of "the cut" in Bergson's *Creative Evolution*—a term that subsequently returns (as *découpage*, or "cutting") in Deleuze and Guattari's *What Is Philosophy?* as well as Deleuze's own *Bergsonism* and *Cinema 1*—as this term will later be mobilized into our proposal for an alternative understanding of the ontology of photography and of the effects it can have. Bergson tells us that "our mind, speculating on it [space] with its own powers alone, cuts out in it, a priori, figures whose properties we determine a priori"[21] and that "while the organized being is cut out from the general mass of matter by his very organization, that is to say naturally, it is our perception which cuts inert matter into distinct bodies."[22] Arguably it is precisely the dialectical relationship between flux and stasis, between duration and the cut that organizes the conceptual and affective universe for us. Bergson seems to be saying as much when he claims that "things are constituted by the instantaneous cut which the understanding practices, at a given moment, on a flux of this kind, and what is mysterious when we compare the cuts together becomes clear when we relate them to the flux."[23]

Although it may initially seem that neither Bergson nor Deleuze are particularly enamored of the practice of the cut, we want to suggest that the cut actually occupies a more ambiguous role in their writings, establishing itself as a fundamental structuring device for their respective yet intertwined philosophies of perception, imaging and temporality. For Deleuze, in a similar way to Bergson, intuition is a "problematizing method" that leads us to undertake "a critique of false problems and the invention of genuine ones."[24] (As stated in chapter 1, this designation of "falseness" with regard to those problems refers to the fact that these are "nonexistent problems," "badly stated questions," or "badly formed composites"[25] but not to any intrinsic moral valorization of them.) Indeed, for Deleuze intuition itself is premised on the act of cutting: it works through "*differentiating* (carvings out and intersections), *temporalizing* (thinking in terms of duration)."[26] So, paradoxically, in order to deal with cuts, intuition needs to make cuts; in order to traverse carvings out and intersections, it must carve out and intersect. This is to say that the only way we can grasp process or time, even if we are to forgo purely *intellectual* understanding and rely primarily on intuition, is through its entanglement with what Bergson calls "solids."

Eadweard Muybridge comes to mind as a photographer who experienced the stoppage of time and who made a memorable cut into the temporal process. We are not referring here to his famous photographs of a galloping horse with all four of its hooves off the ground, as they only confirm what Bergson told us about time and duration through his explication of Zeno's paradox. (Like Muybridge's horse, Zeno's arrow seems motionless during the whole time it is moving. Yet it is only each subsequent point that the horse/the arrow passes through that is motionless; the horse/the arrow never finds itself *in* the point of its course. Our erroneous perception—be it photographic or "immediate"—reduces indivisible movement to a sequence of static states.)[27] Instead, what we have in mind is the moment of arrested passion, suspended perception, and interrupted duration when Muybridge killed his wife's lover: an (at least partly apocryphal) frozen frame that Frampton has described as "*Man raising a pistol and firing.*"[28] We learn as much from Derrida's response to Bernard Stiegler in *Echographies of Television*, when Derrida suggests that "to speak of a technical process, and indeed of its acceleration, mustn't lead us to overlook the fact that this flux, even if it picks up speed, nonetheless passes through *determined* phases and structures." Derrida goes on to admit that what bothers him about the word "process" "is that it is often taken as a pretext for saying: It's a flow, a continuous development; there is *nothing but* process." He then insists: "No, there is not *only* process. Or at least, process always includes stases, states, halts."[29] Though Bergson, and Deleuze after him, would perhaps not be averse to such a pragmatic recognition of the need to stabilize—after all, Bergson himself acknowledges that "the truth is that we change without ceasing, and that the state itself is nothing but change,"[30] which can be seen as a restating of the chicken-and-egg dilemma—it is nevertheless on the flux and duration that both he and Deleuze focus their call to attention. Yet, what if we were to pick up on Bergson's (entertained, engaged and then quickly dismissed)[31] pragmatism and turn it into a critical force? What effect could this have on what we think about duration, temporality, and photographic cuts?

A Categorical Imperative to Make a Cut

We want to suggest (with Bergson and Deleuze, but perhaps also against them) that this call to intuition can be understood as a call for a temporary suspension of the method of knowing—that is, carving—the world. This is not to offer instead a kind of "go with the flow" method or attitude, but rather to postulate the need for what could be termed "differential cutting"—which is another name for "cutting well." Further, if cuts are inevitable and if cut indeed we must, then photography can be seen as a transintuitive practice of *working through the cut,* of re-cutting and re-cising things "for good measure." *It is precisely in the gap between photographs as media objects, and photography as a practice of mediation that aims and fails to capture the passage of time, that an ethical imperative presents itself to us.* This imperative entails a call to make cuts where necessary, while not forgoing the duration of things. Rather than being reduced to a technique for providing false renderings of the world which is ultimately unstable and moving, photography can be said to lend us a helping hand in managing this duration of the world. Yet is this not to invest photography with too much ethical weighting, too much responsibility, and hence too much agency? Are we not expecting it to carry out the work of technology (i.e., *tekhnē*, bringing-forth, or creation) but also of safeguarding the novelty of what it has created?

Karen Barad's notion of "the agential cut" developed in her book *Meeting the Universe Halfway* can provide us with some ideas on how to engage with the agency of matter and how to raise ethical questions in a world which is relational, durational, constantly becoming. Drawing on Niels Bohr's work on quantum physics, Barad develops a philosophical position that remains attentive to "matter's dynamism," recognizing as she does that matter is "an active factor" in any ongoing and further materializations of the world.[32] Her theory is therefore not about what we can do *to* the world and how we can act *on* it, but first of all about how we intra-act *with* it. The concept of intra-action postulates a more dynamic model of emergence of and with the world, whereby "the boundaries and properties of the components of phenomena become determinate" and the particular concepts become meaningful only in an active relation. Intra-action thus points to the inherent performativity of matter, and of the relations it enters into on many scales. To the theories of performativity studied in media and cultural studies via the writings of Austen, Butler, and Derrida, Barad adds the knowledge of physics.[33] Her work deserves attention due to the serious effort she makes to work through science, while not losing sight of philosophical and cultural debates. Driven by a conviction that "scientific theories are capable of providing reliable access to the ontology of the world,"[34] Barad also remains aware of what we could call human singularity (not to be confused with human superiority!) that emerges as a result of her "agential cut." She goes on to argue that "the fact that science 'works' does not mean that we have discovered human-independent facts about nature."[35] The agential cut becomes for her a material-conceptual device, whereby a distinction is introduced between "subject" and "object" within the larger material arrangement. In other words, "the agential cut enacts a resolution *within* the phenomenon of the inherent ontological (and semantic) indeterminacy. . . . Crucially, then, intra-actions enact *agential separability*—the condition of *exteriority-within-phenomena*."[36]

Commenting, after Bohr, on the use of apparatuses in physics experiments, she recognizes that they are not "passive observing instruments; on the contrary, they are productive of (and part of) phenomena."[37] If we apply this to photography, we can see how the camera as a viewing device, the photographic frame both in the viewfinder and as the circumference of a photographic print, the enlarger, the computer, the printer, and the photographer (who, in many instances, such as CCTV or speed cameras, is replaced by the camera eye), are all active agents in the constitution of a photograph. In other words, they are all part of what we understand by photography. Indeed, Barad points out that causal intra-actions within the world "need not involve humans"—although the semiotic designation of a set of these practices and actions as "photography," and the cultural valorization thereof, does.[38] The role of the agential cut, enacted by human and nonhuman agents alike, is to divide photography into photographs, but then to reconnect the latter to its beyond: that is, photographic duration whose stabilizations into artifacts are only ever temporary. (They are also, to cite Bergson, "useful.") As Barad points out, "cuts are part of the phenomena they help produce."[39]

If the cut is indeed "enacted rather than inherent," and if its task is to enact "a resolution,"[40] then we can see how this kind of agential cut has both an ontological and an ethical dimension: it is a causal procedure that performs the division of the world into entities, but it is also an act of decision with regard to the boundaries of those entities. Naturally, only some of these de-cisions will be true *ethical* decisions in the sense of having been made by (or with) a human agent, but the majority of them will have moral consequences. What we mean by a "true ethical decision" here differs from the position on agency in traditional moral philosophy, whereby an ethical decision is made by a transparent, self-contained, liberal subject who is capable of evaluating the available options and making a rational choice from among them. For us, a decision is always to some extent arational, made by the (inhuman) other in me, and necessitating a leap of faith beyond the scope of available options. To answer the question posed at the beginning of this chapter, "What does it mean to cut well?" we thus want to suggest that *a good cut* is *an ethical cut*, whereby an in-cision is also a de-cision. Cutting well therefore means cutting (film, tape, reality) in a way that does not lose sight of the horizon of duration or foreclose on the creative possibility of life enabled by this horizon.

The Vitality of Photography

Again, photography has an important role to play in this process of cut-making because of its proximity to life itself—which, for Bergson as much as for Deleuze, stands for duration, time, and movement. Deleuze writes, "Movement is . . . explained by the insertion of duration into matter." He then goes on to suggest that duration "is called life when it appears in this movement."[41] Significantly, Deleuze also identifies an ongoing process of differentiation occurring within this flux of life, of life becoming other than itself, of cutting itself and into itself. The inherently technological process, whereby *tekhnē* stands for bringing forth and creation, is therefore not just an aspect of human becoming, as it was shown in chapter 1, but also of life itself. Life itself is technical because it is (potentially) creative: it carries within itself propulsion for movement, for change, for the creation of differences. Photography is in a privileged position (even more than cinema perhaps) to capture this flow because it takes on and reveals, instead of concealing, the agential cut which is involved in transforming matter into objects. In this, it *produces* life forms, rather than merely recording them.

Claire Colebrook explains this process of creative becoming in and of life by drawing, interestingly, on the concept of image production, or "imaging." She writes:

> All life, according to Bergson and to Deleuze after him, can be considered as a form of perception or "imaging" where there is not one being that apprehends or represents another being, but two vectors of creativity where one potential for differentiation encounters another and from that potential forms a relatively stable tendency or manner. The intellect is on the one hand a deterritorialization of the organism, for it creates a series of concepts, quantifications, units and measures—including a uniform clock time and a number system—that will allow it to extend its range of action and movement beyond the immediate present, and beyond its own life. On the other hand, the organism is also a *re*territorialization that folds back upon itself, quantifies and measures itself.[42]

Photography's mode is not therefore primarily that of perception; it is rather that of intermittent duration, of incised movement and captured time. It is in this sense that photography can be described as having a lifeness. What we mean by this is something more than the lifelike effect of animated photographs in digital photo frames or on Flickr photostreams. Photography apprehends duration in an instant; we can therefore say that it intuits time by cutting through its solidifications. We see photographically when we intuit things, when we grasp things in an instant, in the blink of an eye. "Intuition gives us fleeting access, which is just like the opening and closing of a shutter," to the movement of things.[43] Drawing on Colebrook's argument, we would thus like to make a second, reverse proposition: all life is photographic. This is to say that life goes beyond and contests representation: it is a creation of images in the most radical sense. Photographic practice as we know it is just one instantiation of this creative process. It is a way of giving form to matter, one that is situated between the three daughters of chaos Deleuze and Guattari wrote about—science, art, and philosophy—in a hybrid category of its own.

However, if indeed all life "can be considered as a form of perception or 'imaging,'" we should perhaps also take heed of Stiegler's proposition outlined in his essay "The Discrete Image" that "*The image in general* does not exist."[44] Drawing on Derrida's deconstruction of the linguistic sign, Stiegler argues that the mental image and what he calls the image-object "are two faces of the same phenomenon." And thus, "just as there is no 'transcendental signified,' there is no mental image in general, no 'transcendental imagery' that would precede the image-object."[45] Also, just as there is no signifier without the signified, there is no photography without the photograph, and no mediation without the media. Arguably, both "the mental image" and "the image-object" come postperception; they are both products of the cuts made into the flow of duration—but cuts enacted in different media (mind vs. emulsion-covered acetate, say). Bergson is not unaware of the incessant and necessary production of such images. However, what his theory of temporality does is turn our attention toward what we could call a missing horizon of reality, and away from an excessive focus on the outcomes of the "fragmentation processes."[46] As shown earlier, Bergson is aware of the practical, useful side of producing these solids, but warns us against losing sight of the incessant transmission of matter on the way. By losing sight of it, says Bergson, we lose sight of the inherent creation of life, that is, its potential to generate things ever new, and thus remain confined to the realm of the "repeateds." In a similar vein, Stiegler acknowledges that "there is neither image nor imagination without memory,"[47] whereby memory is, incidentally, another term for duration and movement, both for Bergson and Deleuze.[48] So the question becomes not so much whether duration or flux is "better" than stasis when it comes to understanding life and ourselves as living beings (as some crude applications of Bergson or Deleuze in media, communications and cultural studies sometimes seem to imply) but rather how to ensure that the agential cut, which is inevitable, is enacted "well." How should one cut in a way that does not lose sight of the horizon of duration or foreclose on the creative possibility of life enabled by this horizon?

We should note here that for Stiegler, there is also no such thing as pure perception (unlike for Bergson, for whom intuition *is* pure perception). Especially in the digital age, our perception is dependent on a number of "intermediary prostheses," argues

Stiegler, with what he calls analogicodigital technology transforming the very way we see the world, and are in the world. What Stiegler calls "the systematic discretization of movement," which he identifies with the emergence of the analogicodigital image[49]—that is, an image that foregrounds rather than conceals the decomposition and synthesis of both light and time—contributes "to the emergence. . . . of another kind of belief and disbelief with respect to what is shown and what happens."[50] He argues: "Digitization . . . introduces manipulation *even into* the *spectrum*, and, by the same token, it *makes phantoms and phantasms indistinct*. Photons become pixels that are in turn reduced to zeros and ones on which discrete calculations can be performed. . . . the *this was* has become essentially *doubtful* when it is digital."[51] Although questions can be, and indeed have been, raised as to whether the analog image did maintain a less problematic relation to the past, the *it was*, the notion of the analogicodigital highlights this uncertain and unstable "knotting" of life, and hence of time, while signaling its discontinuity, or discreteness, *in* time. It also promises us "another intelligence of movement."[52]

Interestingly, in the reiteration of his tribute to Derrida's deconstructive thought in the closing pages of his essay, Stiegler makes a rather Bergsonian point when he says: "life (*anima*—on the side of the mental image) is always *already* cinema (animation—image—object)."[53] This is a position we know well from Deleuze's *Cinema* books as well as Hollis Frampton's idea of "infinite cinema," which, in its focus on the appearances of the world, supposedly precedes the invention of both the camera and the cinematograph. Sekatskiy provides a poignant summary of this evolutionary narrative of cinematic development, a narrative which has gained some currency in film studies over the last two decades: "Moving images are more primordial than still images. We should not be surprised therefore by the speediness with which cinematic effects became embedded in dreams, memory and imagination. If the movie camera is a prostatic [prosthetic?—SK and JZ] device, a contact lens for the imagination, then we must accept that it is hardwired to the very core of our being: its proximity to internal apperception makes the rejection of cinematic fantasies practically impossible."[54] But then, mimicking the scientific jargon of this narrative, Sekatskiy goes on to challenge its inherent logic by drawing, in a truly Bergsonian manner, on our "real-life experience." He writes: "In fact the opposite has occurred. Lazy human reaction has got into the habit of following a guide, of entrusting the 'change of images' to the light-beam of the film projector: this has led to a degeneration of our imagination. It is a different matter in the case of the still photographic camera. Despite a long period of mutual adaptation between the film camera lens and human vision, there is still a marked difference between still photography and human vision."[55]

We are less concerned here with ascertaining the scientific accuracy of this assessment of the status quo (just as we are not as interested in the neuroscientific interpretations of our mind as supposedly cinematic in its "nature"). Instead, we want to treat Sekatskiy's proposition as a philosophical provocation, one that can help us redeem what is being elided in the narratives about the primacy of cinema as both a cultural form and natural imprint, whereby photography is only seen as "a filmic *ersatz*,"[56] and in those rhetoricophilosophical transitions from photography to cinema as presented earlier. This provocation points to the creative potentiality of photography—a dimension that Bergson and Deleuze are in search of in their respective studies of duration, but that they overlook in photography itself. In its physical two-dimensionality, its anchoring in the index (no matter how much of a fantasy this anchoring is), and its existence in the mappable parallel trajectories of art, commerce, and everyday amateur practice, photography becomes a safe zone in which one can take on the chaos of the world. Stored in family albums, on social networking websites such as Flickr and PBase, in data banks and art galleries, it allows us to explore the liquidity of culture without drowning in its fast-moving waters. It is also in this sense that we can talk about photography's vitality, or "lifeness." This is not to suggest that its analog and digital products partake in the processes of self-organization, self-replication, and autonomy in the same way that other forms of so-called artificial life do. It is rather to acknowledge that both in its amateur and professional forms, photography "is capable of carving out new passageways in life, and of life, by moving us, and making us move, in a myriad ways."[57]

The Artist, the Surgeon, and the Philosopher[58]

In the process of worldly becoming, Bergson suggests, "the living body cuts out other bodies from the flow of life."[59] The *Oblique* series[60] by Australian artist Nina Sellars performs the vitality of photography by instituting an operating theater of darkness and light: a hybrid space of surgery and performance that we are invited to enter through nine large window-like images (figure 3.3). In *Oblique*, surgery becomes a theatrical act. It brings back the long gone world of public autopsies and surgeries which used to be held in anatomy and operating theaters for the benefit of medical students but also for the interest and pleasure of wealthy citizens. Yet *Oblique* is more than just an "edutainment" show for those who still remain intrigued by the human body and the processes of its transformation. The artist actually joins the team of surgeons in instantiating a series of cuts and incisions herself, by means of which she brings forth a unique event: a material-visual operation on an apparently disembodied arm and a seemingly displaced ear.

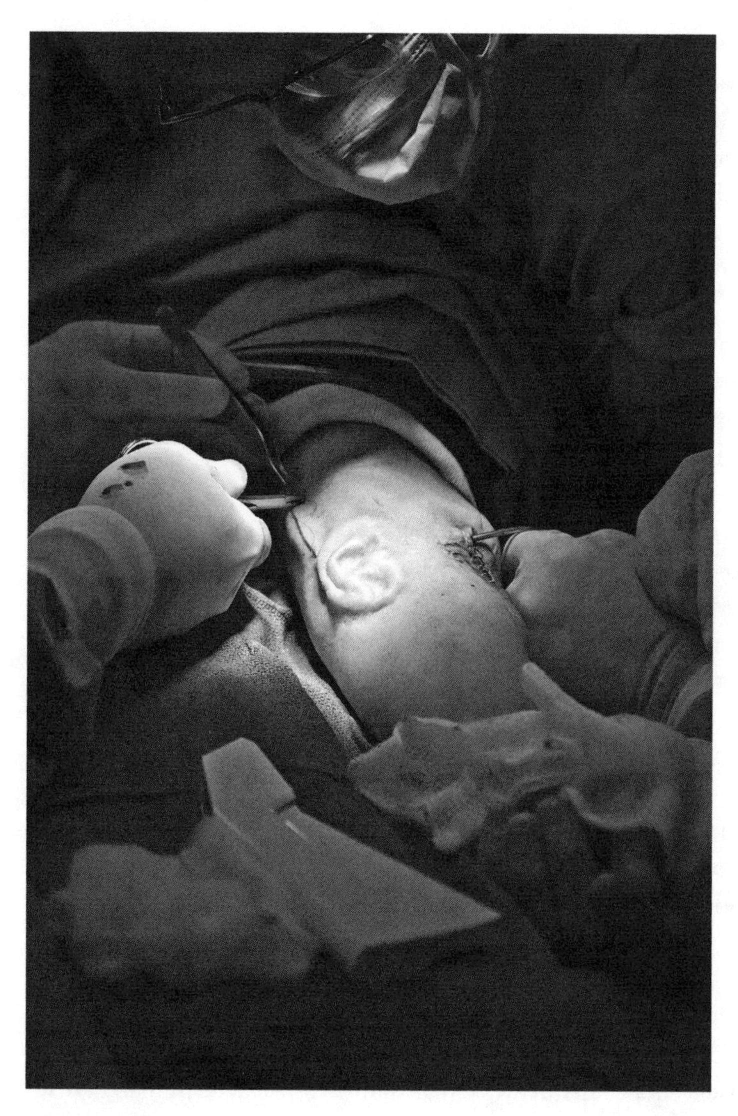

Figure 3.3
Nina Sellars, *Oblique* (2008).

Photography is mobilized on a daily basis to participate in multiple processes of mediation—from those instantiated by the symbolic exchange which is at work once the viewer is faced with the images, to those second-level ones, activated once the photographs enter the media circuit of galleries, publications, TV news, and the Internet. Sellars's photographs allow us to look at bodies in the process of being remade somewhat obliquely, and thus tell a different story from the one offered by the sensationalized and commercialized representations of the open body in mainstream media. Indeed, there is something unique and singular about an encounter with Sellars's unusual (meat) cuts in a gallery space. Herself an experienced prosector as well as a mixed-media artist, Sellars revisits the age-old practice of barber-surgeons, marginal figures within the history of medicine who were nevertheless instrumental in the radical opening up, on both a physical and conceptual level, of the human body.

With *Oblique*, a photographic art project which entails enclosing, freezing, and carving the body into a particular form, Sellars repeats the surgeon's master "cut" with a click of the shutter, which frames and fragments the body spectacle at hand. Arguably, the cut functions as an inherent part of any photographic process, with the photographer freezing, slicing, and carving out a certain instant from the reality surrounding her. Comparing the photographer to the butcher—who, "using only a knife, reduces a raw carcass to edible meat," but does not "*make* the meat, of course, because that was always in the carcass; he makes 'cuts' (dimensionless entities) that section flesh and separate it from the bone"[61]—Frampton seems to suggest that the photographer *makes "cuts" rather than "photographs,"* in-cising the meat of the world in an attempt to give it form. Cutting thus becomes a "component of a practice," to use Victor Burgin's term.[62] It is a creative in-cision that is also a de-cision, because it gives shape to the world; it makes it into this or that. Frampton seems to concur when he says, "The photographic process is *normative.*"[63] Yet the psychic symbolism of art as a metaphorical process of cutting the world, with all the connotations of violence, pain, and pleasure this process entails, frequently remains unacknowledged in many artists' work, as well as in broader debates on indexicality, referentiality, and representation in photography. This material process of cutting has perhaps been more explicitly acknowledged in the practice of film making, both during the shooting stage, when the director shouts "Cut!" to end the filming of a particular scene, and in postproduction, when tape had to be literally cut and spliced together. The digital era's "cut and paste" technique of word processors and other editing programs, such as Apple's Final Cut Pro, has made "the cut" more visible across media platforms and practices: film, photography, graphic design, and writing.

One must of course be careful not to impose moral equivalence between all practices of cutting—for example, slicing someone's face with a knife in a street attack, remodeling a client's face in a cosmetic surgery situation, or metaphorically "cutting" an aspect of reality with a still or film camera. A desire to carve and hence adjust the world according to the "perpetrator's" wish or whim is arguably inherent in all of these gestures. However, even if cutting inevitably entails some degree of violence, our attempt to "cut well" can perhaps be guided by what the philosopher of ethics Emmanuel Levinas has termed "good violence." According to Levinas, violence is constitutive of subjectivity: indeed, the subject can emerge only in response to the intrusion of alterity (i.e., difference). It is "the shock with which the infinite enters my consciousness and makes it responsible"[64] that provides the horizon for an ethical event (even it not guaranteeing that the subject will behave ethically). Such violence is by no means equivalent to violations of identity in war or crime. The reason we can describe it as "good violence" is because "it brings about some good: the responsibility of the subject which is established in this violent act, and which is made ethical—i.e., responding, exposed to the alterity of the other—in its foundation. . . . Our subjectivity is thus always already born in violence, it is a response to what we cannot comprehend, master, or grasp, to what escapes our conceptual powers."[65] The doubling of "the cut" within Sellars's own body of work—from that of the surgeon's scalpel, making incisions in the arm which functions as a focal point for the images, to that of the artist's camera, carving out a moment in space and time while also cutting out this particular operating scene for us in a certain way so that all we can see is an arm, a few pairs of hands and a shaft of light—creates theatrical tension between medical and artistic intervention. Forcing an inevitable question—"Why is someone having an ear constructed on their arm?"—it also introduces a gap between necessity and ornamentation, between lack and excess.

We can imagine Sellars's *Oblique* to be an amplified restaging of Rembrandt's famous *The Anatomy Lesson of Dr. Nicolaes Tulp*: the chiaroscuro color template and the anatomic curiosity of the photographs easily encourage such a comparison. But there is another, more fundamental level at which Sellars puts her "cut" to work. In Rembrandt, the audience—gathered around the operating table to examine the muscle structure on the cadaver—consists of darkly clad gentlemen, curiously poring over the body specimen. There are no cutting instruments, only an open book, as there is going to be no dissection as such: the cadaver will have been prepared in advance for the "lesson."[66] The audience is thus participating not only in a hands-on anatomy class but also in a metaphysical moment of literally trying to grasp "the other side"—to touch death with their hands. Something very different happens in *Oblique*. The close-up photographic technique cuts out not only the audience that may or may not have been gathered around the operating table on which someone is having an ear implanted on their arm—a rather unusual bodily intervention that implicitly asks for an audience because it is in itself a form of spectacle. It also excises the surgeons performing the procedure, turning the operation into an other-bodied dance of hands and arms (figure 3.4). The audience, of course, does not disappear from this spectacle altogether. In the age of makeover TV and ubiquitous plastic surgery—ABC's *Extreme Makeover*, FX's *Nip/Tuck*—the spectators are already part of the picture, regardless of whether they are represented in the actual photographs. Indeed, viewers of mediatized body makeovers are not just watching the transformation of others but are also themselves actively taking on their bodily wounds and corporeal metamorphoses. Such an audience is almost beyond representation: it far exceeds the select few—be it the attendees of the anatomy lesson in Rembrandt's painting or regular visitors to art galleries—to include all the transnational media viewers, exposed to twenty-first century anatomy lessons via franchised shows that turn the surgeon's cut into popular entertainment.[67]

In psychoanalysis, the cut symbolizes castration, which Slavoj Žižek in his interpretation of contemporary bodily modification practices translates into a desire to disentangle oneself from authoritarian ties imposed by what he calls the big Other, and an effort to reestablish a certain kind of individualism outside the constraints of the symbolic order.[68] Sellars's project literally cuts across this fantasy of individualization. What Sellars therefore offers us with *Oblique* is a series of TV screens that are subversively hung as mirrors. Ron Burnett suggests that "viewing is about the desire to enter into the screen and become a part of the images and to experience stories from within the settings made possible by technology."[69] Watching the surgery, we are also getting a glimpse into our own desire for the cut of the other, for his or her suffering, but also for the ultimate transformation and closure that sews up the cut and heals the (psychic and real) wound. The visible surgical tools and the implied camera equipment function as cutting devices for a self on the way to individualization via self-incision and self-revelation. The cut is thus always agential, even if the agent is multiple and not even human.

In the early 1990s, French artist Orlan broke the taboo against the public opening up of the body with her plastic surgery project, *The Reincarnation of Saint Orlan*. She staged her operations in art galleries, with the camera capturing every cut to her face, the detachment of the skin from her head, the slow coming apart of her "mask." Yet Orlan's performances have now been overshadowed by the far gorier and far more detailed exposition of the surgeon's cut in TV makeover shows. What is more, the

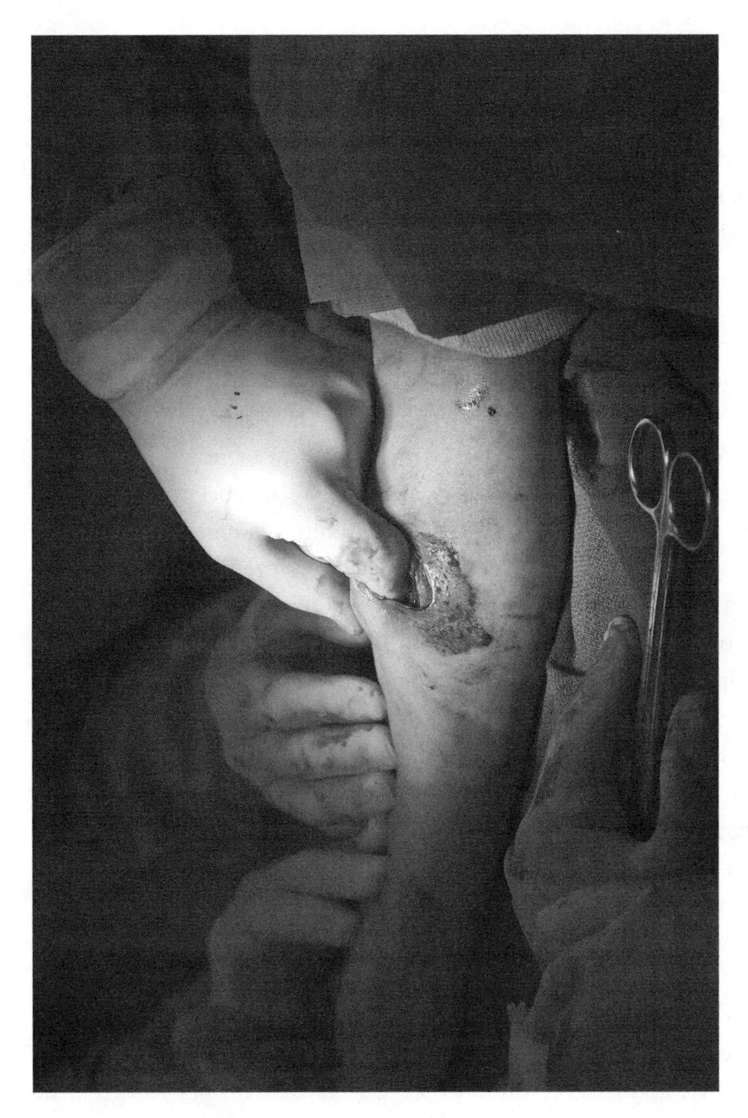

Figure 3.4
Nina Sellars, *Oblique* (2008).

radicality of Orlan's project has been overcome by reality TV's reduction of "the cut" to a mere stage on the way to "the stitch," that is, a supposedly more accurate representation, offering a fantasy of arriving at the fullness of one's being. Sellars, in turn, seems to be embracing "the cut" as a positive condition of one's insertion into the world. The way she casts light and shadow in her photographs does allow us to see an open body, but it also prompts us to look obliquely at the processes of its constitution. Rather than trying to overcome or suture the wound all too quickly, her images dwell on it, repeatedly, obsessively. In this way, they foreground what Parveen Adams has called "the emptiness of the image, not the triumph of completeness that the dominion of the image seeks to induce."[70]

We could therefore suggest that Sellars's project involves the staging of an anatomy lesson for the media age. It is a lesson in which the processes of image production have become much more mechanical, and in which agency has been redistributed among different, human and nonhuman agents—including cameras, large-format printers but also cannulae, sutures, and scissors. *Oblique* therefore provides a glimpse into the originary technicity of the human, where *tekhnē* actively brings forth humanity, rather than being only a promise or a threat to it. Significantly, in Sellars's anatomy class, the surgery is distinctly direction-less. Unable to trace the master hand of the surgeon, we are exposed instead to the medusa-like network of cyborg-like, featureless, latex-gloved hands. As in Rembrandt's *Anatomy Lesson*, the body is the center of each image here, but what we are looking at with *Oblique* is a disassembled body, a body with organs in wrong places, an arm with an ear. Perhaps what these images actually depict is some sort of multilimbed mutant surgeon—a self-mutilator who is folding upon himself in a bizarre act of self-creation (figure 3.5).

The rich visual connotations opened up by the *Oblique* images are nevertheless anchored by the exhibition's subtitle. We learn from it that these are in fact images from Stelarc's *Extra Ear* project, in which this well-known Australian performance artist had an ear constructed on his arm, with a view to exploring what he terms "alternate bodily architectures." By using this performance as raw material for her work, Sellars playfully mobilizes the strategy of remediation: she is trying to "achieve immediacy by ignoring or denying the presence of the medium and the act of mediation"[71] on the level of the photographs' content, while also foregrounding the mediation processes via their screen-like framing. In doing all this, she draws on the cinematic technique of montage, remediating a classical painting that she then multiplies into a kind of film strip. Even if we agree that Sellars's work is a way of holding up a mirror to our own narcissism, our makeover culture-fuelled desire to see "beyond the surface" of the other, or a screen for the projection of our fantasy for completeness, her project

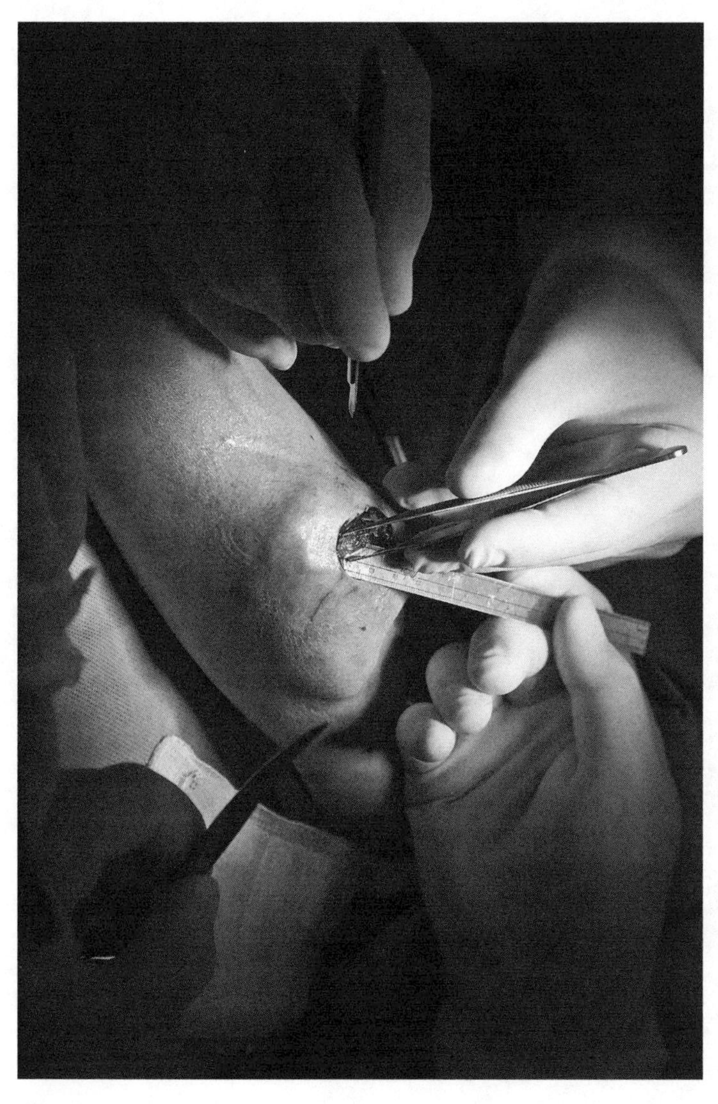

Figure 3.5
Nina Sellars, *Oblique* (2008).

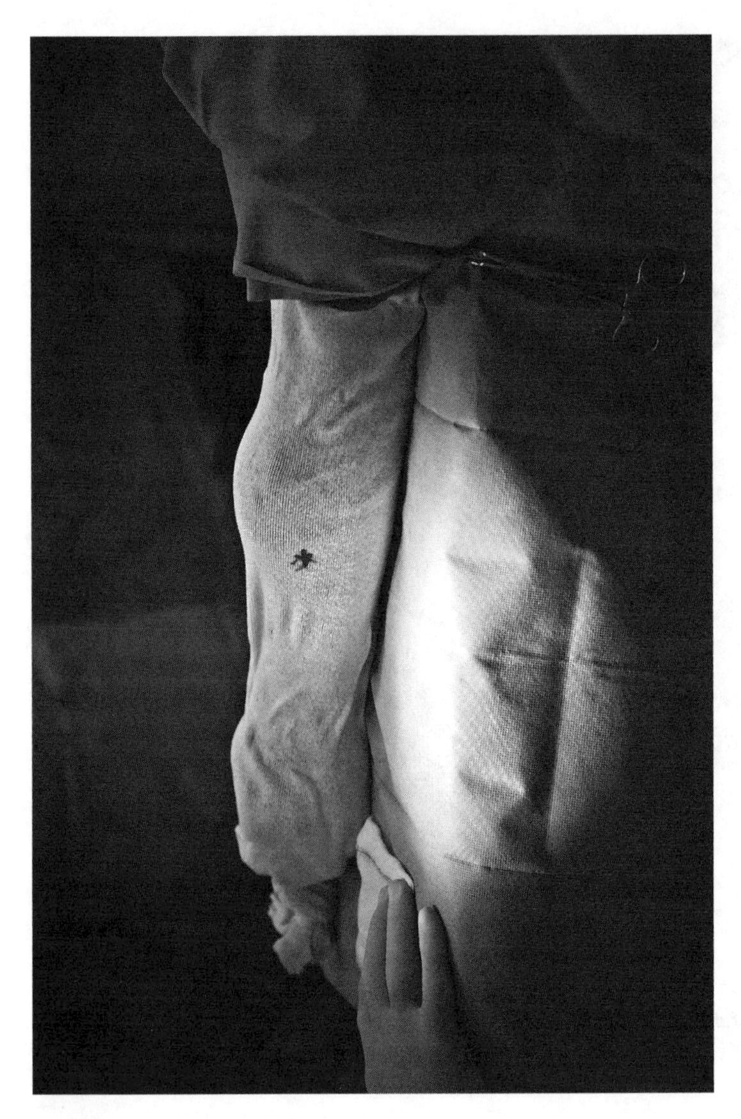

Figure 3.6
Nina Sellars, *Oblique* (2008).

also resists and undoes this tendency toward narcissistic incorporation. In the matte texture of their paper, the tenebrist tonality of their surface and the materiality of their box frame, Sellars's photographs possess a nonhuman agency of their own, which they unwittingly activate to take away what Victor Burgin calls "our command of the scene." Burgin writes: "To remain long with a single image is to risk the loss of our imaginary command of the look, to relinquish it to that absent other to whom it belongs by right—the camera. The image then no longer receives our look, reassuring us of our founding centrality, it rather, as it were, avoids our gaze, confirming its allegiance to the other."[72]

In allowing ourselves to witness Sellars's surgery scene, by being exposed to the gaping hole in the arm, the spectators are participating in their own undoing, in letting the undoing happen to them, in seeing themselves as part of mediation. The images play with metaphysical depth, promising a closure they cannot ultimately offer. This is where their victory as images lies but it is also a moment when the look can return to the viewer—who, as a reward for giving up on the desire to master it, and to stop the passage of time for ever, will be presented with his or her own temporary restoration. The pleasure of looking at *Oblique* is thus ultimately the pleasure of survival, of getting over "the cut" (figure 3.6). It combines the horror evoked by the object of desire—the other's body, the ear, immortality itself—with the narcissistic reconfirmation of the self that has not been cut open, and that can breathe a sigh of relief while saying: "This is not me, I haven't been cut." But it also provides a safe entry point for consciously taking on the self's experience of mediation and fragmentation, with any potential psychosis being transformed into art. Jacques-Alain Miller suggests that any enunciation of the "I" position is always at the same time an act of suturing, of patching over discursive and corporeal instability, somehow against all odds. Naturally, art is not the only terrain on which such a remedial work of recognizing and working through the inherent instability of the "I" can take place. Yet stagings of an open and wounded body in projects such as Sellars's *Oblique* and Orlan's *Reincarnation* offer a safe platform for temporarily taking on and reliving someone else's corporeal and psychic wounds, while learning to perform the work of self-suturing. Devoid of the sugary redemptive promise of TV makeover shows, photography and imaging of this kind—presented within the relatively safe confines of the gallery, the art book, or art video footage—cuts well both through our narcissism and our anxiety, without leaving us out there to bleed to death.

Interlude: I Don't Go to the Movies[1]

A desire to carve the world into little pieces and hold it in her hand is every cinematographer's dirty little secret.

A similar desire lies at the heart of every love affair.

The price of seduction is that the secret must not be broken at the risk of the story's falling into banality.

Yet love has no time for these secrets and neither does cinema.

Stealing his traces, I cut, I shoot, I murder.
Then I change my mind.

I look away and move on.
Simply and unequivocally, he is now dead to me.

1. Joanna Zylinska, *I Don't Go to the Movies* (2009), selected images.

4 Home, Sweet Intelligent Home

Transformative Mediation of Technological Agency

One of the key issues we aim to address in this chapter is some of the ways in which technoscientific discourses—in particular, those related to developments in ubiquitous computing which are centered on the so-called smart home—purport to represent and, at the same time, contribute to bringing about certain entities and notions. We will show how *technoscience*—which for us in this chapter stands for the discourses, methods, philosophies, and practices within the electronics and computing industry—posits the self as an entity which is more or less fixed, universal, yet remaining in need of some reinforcement within the home, with the latter defined as an environment of "new," intelligent media. We will also show how the electronics and computing industry thus contributes to the transformative mediation of subjects, the social, and of biopolitical life itself *after* new media—that is, after the era of the stand-alone *computer*—into the network of embedded and distributed *computing*. We will investigate how much of a change this supposed technological shift actually offers and to what extent the industry seeks to disengage the transformations it contributes to bringing about.

The broader aim of this investigation, as explained before, is to clarify and extend Jay Bolter and Richard Grusin's understanding of mediation as a multiagential process that incorporates media technologies and users (as well as organizations, institutions, economies, etc.). In order to do so, we are turning to Bergson's intuitive method, which allows us to take things apart in order to see, or see better, how they fit together. So, rather than collapse mediation *as a process* with media *forms*, or regard mediation as purely a function of media objects such as photographs or film (the way we have argued that Bolter and Grusin do, in line with mainstream representationalist media theory), we separate the two terms, albeit contingently. We do so in order to try and establish that "media" and "mediation" are differences in kind, not degree—differences that are rendered irreducible by the emphasis they respectively place on space and time, on representation and performativity.

We understand performativity to refer to statements, beliefs, images, and stories that contribute, discursively and materially, to bringing about the events that they describe. In the context of agential contribution to the emergence of a phenomenon, we suggest that performativity works as a *critique of causation*, of the linear cause-and-effect determinism, or "mechanism," that Bergson critiqued along with teleology, or thinking in finalist terms. Mechanism and finality are for Bergson "two ready-made garments that our understanding puts at our disposal."[1] Ubiquitous as they are in our attempts to understand life, these two approaches fail to capture life's dynamic vitality, reducing it to a predefined trajectory and hence foreclosing on the possibility of acknowledging and recognizing anything truly unexpected and new. Even though discussions of performativity are already well rehearsed in media, communications, and cultural studies, we suggest it is important to return to the notion of performativity in the context of discourses around new media technologies in general and ubiquitous computing in particular in order to counteract the deterministic, finalist tendencies that underpin many academic and mainstream debates on technology. Indeed, determinism and notions of progress still feature rather prominently in the dominant conceptions of technology, not least by producing their own counterforce—a humanist answer to technicism that results in historical reductionism and in the foreclosure of any radical technological change (an approach that, for example, seeks to reduce the Internet to the telegraph with a supersized memory).[2] Such counterforces or reactions are arguably no less flawed than the positions they oppose. Yet the question of technological agency, often staged as a "debate" between Marshall McLuhan and Raymond Williams within media, communications, and cultural studies, is complex, indeterminable, and by no means resolvable by taking recourse to the supposed autonomy of either machines or their human "users." In other words, there are no obvious winners or culprits in the "whodunit" narrative about technological change.

Indeed, our argument in this chapter is that *the question of technological agency is most commonly explored and enacted through the interplay of autonomy and relationality*. Contemporary technoscience, as we will show, has a stake in this interplay. To a large extent, technoscientific articulations of technological agency have typically adopted an explicitly posthumanist tenor, concerning themselves with establishing alternative or even successor intelligent life forms within hardware, software, and wetware.[3] However, some more recent developments in ubiquitous computing have made an explicit recourse to humanism (which, arguably, already underpinned earlier forms of supposedly posthumanist technoscience). Ubiquitous computing is thus defined as seemingly centered on "us," on our very human needs and our domestic environment. In this way, it simultaneously suggests and conceals a more transformative potential that is inherent in this particular technology. We want to argue in what follows that in foregrounding technological transformation, ubiquitous computing belies, or even absorbs, the philosophy and politics of *actual* change.[4]

Many forms of ubiquitous computing such as smart homes and wearable technologies are being developed at present. But in as far as they are inherently science-fictional, that is, performatively produced by the confluence of science and fiction, they highlight what for us is an underdeveloped aspect of contemporary thinking about the supposed autonomy or relationality of things. This is its methodological, or "how-to," aspect, which stands for both a critique of the current electronics and computing industry and a need to develop, with a nod to Bergson as well as Rosi Braidotti[5]—*a new mode of address* with regard to relationality between humans and machines. This new mode of address would be able to suggest what cannot be expressed and to match the dynamism of its presumed object. If, as the work of both Bergson and Braidotti testifies, it is not possible to capture or express life as it is lived, or to grasp technoscientific and other changes as they occur (since things have to be seen *in* and *as* relations), it is certainly possible to *enact* the vitality we intuitively apprehend and otherwise intellectually disengage. One way to do this is by combining, for example, images with ideas, stories with theories. For Karen Barad in *Meeting the Universe Halfway*, such a methodological maneuver would be tantamount to closing the ontological gap between representations and that which they purport to represent.

Makeover or Metamorphosis?

Our study of mediation and (supposedly) intelligent media in this chapter runs alongside our investigation of the current global financial crisis; of the boundless potentialities of life, the universe, and "everything" that stem from physicists' attempts to recreate the Big Bang at CERN discussed in chapter 2; and of the makeover and metamorphosis of the self in cosmetic, reconstructive, and face transplant surgery outlined in chapter 5. With this, we want to argue that *futuristic discourses of the smart home are allied to a wider culture of makeover and metamorphosis*, prefiguring as they do a sense of what we could become in the cosmic and cosmetic future, and what could become of us in the more immediate economic sense. In as far as we can associate the financial crisis with subprime mortgages and an inflating and deflating housing bubble, then, as

Joshua Hanan points out in "Home Is Where the Capital Is," the housing bubble saw the role of the home transform from a roof over one's head to a site of convergence between self-identity, technology, and capital.[6] This altered—or, rather, intensified—role of the home (which capitalism has posited as one of the potential sources of investment) corresponds to the neoliberal mode of governance in which control is more flexible, fluid, and, as we will show, rather friendly. There are correspondences here between a moral obligation to remodel one's financial portfolio, one's home, one's body, and one's face, exercised by the news about financial catastrophes as well as "makeover TV" reality shows that show house remodeling and cosmetic surgery live.

Our principal interest in this chapter is in the role technology plays within this converged, or, as we prefer to call it, *remediated* environment. We want to look at some of the ways in which technology contributes to the coming together and repurposing of self-identity and capital. In order to do this, we will look at how developments in ubiquitous computing contribute to the emergence of a marketized self, which we could describe as a "data subject."[7] We will also explore whether the self is merely renewed, or "made over," in the process, or whether a more radical metamorphosis is perhaps enabled by ubiquitous computing, with its promissory-sounding "Ambient Intelligence."

Ambient Intelligence is both a brand name for ubiquitous computing and a development of the project associated with Xerox PARC (the Palo Alto Research Center) and the work of Mark Weiser.[8] If the (as yet unfulfilled) idea behind ubiquitous computing involves replacing obtrusive and demanding *computers* with unobtrusive and user-friendly, that is, flexible and fluid, networked *computing*, then Ambient Intelligence, the branding of ubiquitous computing by Phillips researchers, places more emphasis on the environmental and social elements of human and computer interaction that are already in play.[9] Indeed, it claims to offer the user "a natural interaction" with "dissolved electronics."[10] An intelligent environment is conceived of as one that is capable of recognizing the people who inhabit it and then adapting itself to them, learning from their behavior, and even exhibiting some form of emotion. In an Ambient Intelligence world, people are allegedly more likely to be at leisure than at work, to feel at home rather than like they are working in an office. They will not be engaging with technologies designed to enhance their productivity but rather with those that enhance their lives and even anticipate their needs. Such industry-led claims seek to disarticulate and deny the entanglement of life and capital that the Ambient Intelligent industry itself contributes to bringing about. This disentanglement is to be achieved by drawing on seemingly human-centered notions of invisibility, agency, relationality, affect, and intimacy.

Invisibility in the Future Home

Ambient Intelligence as a form of computing without computers[11] is, or will be, *embedded* and *concealed* within the objects and materials of everyday life. This recourse to the everyday, the quotidian, might be regarded as a marker of the significant downscaling of the more hubristic projects of Artificial Intelligence (AI) and Artificial Life (ALife)—the building of intelligent robots and creation of algorithms that can breed and evolve and that constitute life forms according to specific technoscientific criteria.[12] But although the technology of Ambient Intelligence (AmI) seeks to display emergent properties "similar to those in biological and neural systems"—in other words, similar to those manifested in the discourses of AI[13] and ALife—it also contains the hidden potential for increased "surveillance, regulation and rationalisation."[14] Fiona Allon recognizes this potential in a futuristic discourse that is marked, partly by virtue of its investment in invisibility, by a somewhat nostalgic vision of home and family. Insofar as computation will be embedded and concealed within the domestic environment, then, as Stefano Marzano suggests, "the living space of the future could actually look more like that of the past than that of today."[15] Our homes will appear to be relatively free of technology; they will look more like the *actual* home of the 1950s than the future home of the 1950s. Nostalgia, of course, is a feature of futurism in general. As futurists point to the horizon and predict how our lives will be transformed by technology, they invariably draw on "socially conservative images and the traditional values of an apparently simpler age."[16] This may help to explain at least one striking similarity between the Monsanto House of the Future (circa 1957) and Microsoft's "Future Home."

The Monsanto House of the Future was a walkthrough attraction at Disneyland from 1957 to 1967. It advertised new manmade building materials, notably plastics (melamine worktops, vinyl floors, etc.) and featured domestic technologies such as the dishwasher, the microwave, the climate-control system, and the telephone intercom, complete with closed circuit television. It is still possible, courtesy of YouTube,[17] to watch these then futuristic gadgets being operated by a wife and mother—one of twenty million visitors[18]—who dreams, in color, of her perfect day. In between her husband leaving for and returning from work (he operates the stereo and the entertainment system, while she and her daughter-helper deal with everything concerned with household management), she has "fun" in the kitchen, which has "a place for everything and at your fingertips." The fun, apparently, "is in making the most of the ultimate in kitchen convenience and efficiency," features represented by the extensive use of plastic utensils and the consumption of "irradiated food." As our happy housewife leaves the kitchen and moves through the boy's and girl's bedrooms, her outfits change. She eventually loses the pinafore, exchanging it for evening wear and earrings just before her husband returns to occupy the living room. At the end of the day, she strokes the "decorative plastic laminated safety glass" in the master bedroom and wakes up, in black and white, to her present reality in the Disneyland attraction: "maybe, someday."[19]

If we compare the 1950s house of the future with that of the present day, how much has really changed, both on the level of reality and fantasy? To help us answer this question, we should look "Inside Microsoft's 'FutureHouse,'" also courtesy of YouTube. In the video, a woman called Janet introduces us to her kitchen PC, which is "easy to clean" and helps her with the cooking. When she places a bag of flour on her melamine worktop, it asks her (in a female voice) whether she would like some assistance. "Yes," replies Janet. Using interactive technology, speech recognition, and machine learning, the worktop has developed some idea about what this particular housewife likes to cook and duly offers some suggestions—displayed, in text format, on the worktop surface. As well as telling her how much dough to use for her focaccia, the worktop reminds Janet to take her medication. A camera installed in the ceiling reads the label on the bottle and confirms that this is the correct dose. Janet moves to the dining room, where the table functions as another computational interface, enabling her to transfer information from a business card to her mobile phone just by placing both items on the surface. There is interactive digital wallpaper in her teenager's bedroom so that the said teenager "can throw things" and play games. There is also a mirror that displays information about the weather, reads tags on one's clothing, and dispenses fashion advice.[20]

As we can see, some rather predictable gender patterns are embedded in the majority of technofuturist visions. Danny Briere and Pat Hurley's "virtual day in a fictitious smart home" as outlined in their book, *Smart Homes for Dummies*,[21] also stars a woman. She maintains a punishing schedule of predominantly domestic activities from "first light" to midnight without once leaving the house. She has a spouse, several children, and pets. She gets them all up in the morning and puts them all to bed at night. In between, she operates the "home network," which controls everything from music to lighting to heat. It does not, however, make breakfast, lunch, or dinner or deliver coffee to the spouse while he is reading a printout of news headlines and personal stock.[22] The woman works and eats lunch in the sunroom, works out in a private gym ("when you enter, you announce yourself to your voice-activated home-automation system, and it automatically sets the music and other environmental settings to your previously defined preferences"),[23] conducts a video conference in the basement ("while in the basement, you call up your home-control system and start the roast cooking in the oven"),[24] clears up after dinner, monitors her family using "the picture-in-picture (PIP) capability on your TV set,"[25] reads her electronic book and presumably passes out, sleeping peacefully in the knowledge that "all night long, your home controller and its various sensors keep an eye on everything for you."[26]

The American feminist writer Betty Friedan was criticized for centering her concept of the feminine mystique—the idea of the happy housewife and mother—on white, middle-class women at the expense of other social groups with very different domestic experiences, but such a model woman still appears to be evoked in contemporary visions of the technologically assisted homemaker. These future-oriented visions are normative and strangely regressive, relocating women (back) into nuclear heterosexual families and into the home—in which, once again, "housewifery expands to fill the time available."[27] It was while pondering this phenomenon that Friedan conceived of the paradox of the feminine mystique: "that it emerged to glorify woman's role as housewife at the very moment when the barriers to her full participation in society were lowered, at the very moment when science and education and her own ingenuity made it possible for a woman to be both wife and mother and to take an active part in the world outside the home."[28] If Friedan's work preceded and helped to produce so-called second-wave feminism, why are we hearing echoes of the feminine mystique in a supposedly postfeminist era? Critics of postfeminism would suggest that feminism's agenda is weakened when it is assumed to be already achieved. This, as Lionel Shriver points out in her introduction to the 2010 edition of Friedan's book, "masks a host of stubborn realities."[29] Similarly, Angela McRobbie rejects third-wave and postfeminist teleology as a kind of conceptual and political forgetting, even while she remains alert to the makeover as well as metamorphosis of feminism within neoliberal forms of governance.[30] We would argue, therefore, that it is important to attend to the not-so-subtle reappearance, and the supposedly invisible manifestation, of regressive gender ideologies in the increasingly proliferating visions of the future home, even if that manifestation is largely incidental and associative, part of a more general nostalgia for a time when the technohouse of the future first emerged.[31] It was a time when, as the popular TV series *Mad Men* so accurately depicts, the postwar feminine ideal of housewife and mother, as incarnated by Betty Draper, was in play alongside the real possibilities opening up for women in the workplace (*Mad Men*'s first female copywriter, Peggy Olson) and in the context of an emerging consumer culture that divided and discriminated among consumers on the basis of a range of factors, including gender, race, and, most of all, economic status. Then as now, the vision of what our homes will be like and how we will live in them was industry led, and the equivalent of Monsanto's 3D advertisement for plastics is Microsoft's plug for the extension of personal computing into the objects and materials of everyday life.

The Door Strikes Back, or the Agency of Objects

Stefano Marzano predicts that, "As technology becomes hidden within these static, unintelligent objects, they will become subjects, active and intelligent actors in our environment."[32] Regardless, for the moment, of any actual topic of conversation, what does it mean to be able to converse with a worktop, mirror, refrigerator, or toilet? What does it mean to have them respond to us in any other than a purely functional way? What happens when their functionality changes, develops, and displays signs of autonomy? These are the sort of questions concerning machine agency that have long been at the heart of the technosciences themselves as well as of technoscience studies[33] and, of course, of the technological imaginary. Arthur C. Clarke's clever, chess-playing computer HAL 9000,[34] who turned nasty on astronaut Dave when Dave just wanted to get back inside the spaceship, comes to mind here:

"Open the doors, HAL."
"I'm sorry, Dave. I'm afraid I can't do that."

After murdering everyone else aboard, including poor Frank, HAL seemed to express some sadness: "It's a shame about Frank,"[35] he declared. Many of us can probably also remember Brian Aldiss's attempt to create a more loveable life-form in the shape of David,[36] an artificial Pinocchio who honestly believed he was a real boy. Most beguiling perhaps is Douglas Adam's depressed robot, Marvin ("I'm not getting you down at all, am I?"),[37] who reluctantly engaged with space travelers and parodied obsequious artifacts: "The irony circuits cut into his voice modulator as he mimicked the style of the sales brochure. 'All the doors in this spaceship have a cheerful and sunny disposition. It is their pleasure to open for you, and their satisfaction to close again with the knowledge of a job well done.'"[38]

Not least because of its performative role in technosciences such as AI and ALife,[39] the majority of theorists and practitioners agree on the need to take science fiction seriously. It *is* actually worth considering whether our doors will be like those of Arthur C. Clarke, Douglas Adams, or, as Adam Greenfield fears and predicts, rather more like those of Philip K. Dick:

The door refused to open. It said, "Five cents, please."

He searched his pockets. No more coins; nothing. "I'll pay you tomorrow," he told the door. Again he tried the knob. Again it remained locked tight. "What I pay you," he informed it, "is in the nature of a gratuity; I don't *have* to pay you."

"I think otherwise," the door said. "Look in the purchase contract you signed when you bought this conapt."

In his desk drawer he found the contract; since signing it he had found it necessary to refer to the document many times. Sure enough; payment to his door for opening and shutting constituted a mandatory fee. Not a tip.

"You discover I'm right," the door said. It sounded smug.

From the drawer beside the sink Joe Chip got a stainless steel knife; with it he began systematically to unscrew the bolt assembly of his apt's money-gulping door.

"I'll sue you," the door said as the first screw fell out.

Joe Chip said, "I've never been sued by a door. But I guess I can live through it."[40]

Although the opacity of AmI technology does not immediately lead the nonspecialist to questions about the agency and accountability of such technology, the introduction of speech-based autonomous agents into AmI devices that may take the form of software or hardware brings such questions to the fore. In the contexts of AI and ALife, the question of agency has typically been pursued by focusing on their supposed agents' adoption of animal or humanoid forms. However, we are now faced with what Mark Andrejevic calls "appliance animism"[41] and the actual prospect of a smart toilet: "an instrumented toilet capable of testing the urine for sugar concentration, as well as registering a user's pulse, blood pressure, and body fat."[42] Matsushita Electronic's prototype "is almost certainly a new frontier for biotelemetry," as the user can have his or her "data" sent directly to a doctor, courtesy of the toilet's built-in Internet connection.[43]

How will we then relate to animated appliances that may communicate directly with us, or with each other, or with others about us? The opportunities for the sort of paranoia and projection that have characterized human-machine relations in fiction as much as in real life will surely increase. In *Human–Machine Reconfigurations*, Lucy Suchman reminds us that our fears and desires concerning the technological other are organized around the belief—often stubbornly held and needing continual reinforcement—in our own autonomy and our essential distinction from machines. This belief is aided and abetted by the "tropes of liveness" that animate machinic agents, thus creating the illusion that such agents move and act alone.[44] The discourse of distributed agency and human-machine relationality as it is presented in an industry context seems rather disingenuous; it is nothing more than a form of conceptual containment. With this, the "project of designing intelligent artifacts" is certainly consistent, as Suchman argues, with the Western philosophical tradition that treats autonomy, not relatedness, "as the mark of humanity."[45]

Human-Machine Relationality

The Ambient Intelligence project incorporates the kind of embodied and situated agents[46] that used to preoccupy ALife researchers and theorists and that subsequently led to the critical reappraisal (both inside and outside the field) of humanism in technoscience.[47] The AmI project appears to have finally found an answer to the question as to whether, or to what extent, artifacts themselves can be considered intelligent or alive, and therefore whether they have subject status or even cyborg rights.[48] It has arrived at this answer by shifting its focus from the technological objects or artifacts themselves to their relational qualities. The key question that emerges for us here is the actual meaning of the term "relationality" in this context.

In their chapter "Intimate Media: Emotional Needs and Ambient Intelligence" that forms a contribution to *The New Everyday: Views on Ambient Intelligence,* John Cass and colleagues describe what is at stake, from an industry point of view, in turning a house into a home. A home, for them, is a space in which individual memory and history is externalized in "complex collections of artifacts: books, images, letters, souvenirs, etc."[49] These artifacts acquire the status of an "extended self" and act as triggers in a "cycle of self-reinforcement."[50] Relationships, demonstrated "through collections of objects and photographs in their space," are merely part of this same cycle of self-reinforcement.[51] In another piece in the same volume, Marzano confirms that in this new manifestation of AI, all of the attention will be on us, rather than on machines that can talk, think, play chess, and attempt to lock us out of our spaceships. The technology of AmI is thus positioned as chastened, humble, inconspicuous, and servile. It will supposedly "recognize us, notice our habits, learn our likes and dislikes, and adapt its behavior and the services it offers us accordingly."[52]

Suchman connects this expected behavior to the broader logic of the service economy and recognizes its consistence with the main tenets of neoliberalism as well as with the traditional master-slave dynamics in historical relations between humans and machines. There is, she points out, "a deep and enduring ambivalence" at the heart of these relations, as the agent that serves you is always capable of turning (back) into the one that controls you.[53] This, arguably paranoid, image of master-slave relations is precisely refigured and even extended in AmI, which envisages artifacts and devices that "will also show their intelligence by communicating among themselves."[54] In AmI, such machine-machine relations thus enter the cycle of self-reinforcement in which the human self is regarded as paranoid and insecure, and in which reinforcement answers a supposedly basic (human) need for security, integrity, and autonomy.

Greenfield argues that, in the context of AmI (or, as he puts it, "everyware") "relational" refers to the fact "that values stored in one database can be matched against those from another, to produce a more richly textured high-level picture than either could have done alone."[55] That picture might be of you, and it might be produced for marketing or surveillance purposes. Equally, it might be of the variables affecting the price of your Starbucks coffee.[56] Machine-machine relations create what Andrejevic calls the "digital enclosure," while at the same time enabling it to function.[57] As far as Andrejevic is concerned, all of the interactions that take place within such an enclosure "are shaped by asymmetrical power relations" and the dominant interests of commerce and/or policing.[58] This is why it is important to offer *a critical examination of the industry-led discourse of Ambient Intelligence that will recognize the asymmetries such a discourse seeks to conceal—while also avoiding any tendency to reproduce the very paranoia and technophobia on which it is based.*

The asymmetries, or "dissymmetries," that underline human-machine relations, might be obscured and elided not only in an industry discourse of relationality but also in a *critical* discourse on it, as argued by Suchman. A sense of relationality as the mutual constitution of humans and machines was originally developed, as Suchman points out, in the context of technoscience studies, as a "corrective to the entrenched Euro-American view of humans and machines as autonomous, integral entities that must somehow be brought back together and made to interact."[59] Yet no matter how valuable this concept is as a critique and an intervention, the sense of relationality as mutual constitution should not be equated, for Suchman, either with the establishment of fixed ontologies or with the idea of relational symmetries. Instead, she argues that "we need a rearticulation of asymmetry, or more impartially perhaps, dissymmetry, that somehow retains the recognition of hybrids, cyborgs, and quasi-objects made visible through technoscience studies, while simultaneously recovering certain subject-object positionings—even orderings—among persons and artifacts and their consequences."[60] There is thus a clear political and ethical dimension to such dissymmetry for Suchman.

Picking up on this point, we will show in this chapter how a discourse oriented around animated artifacts repositions the human self as an object within the neoliberal economic and political rationality. The emergent self, in the process of *becoming data*, takes part in a significant reordering of the values that have been traditionally associated with subjectivity and citizenship, even while those same values appear to be reinscribed and reinforced in the process. AmI is therefore rather more hubristic than it appears to be, as the subject it is producing is a metamorphic one being retroactively made over. Another way of putting this would be to ask why AmI makes over, or even "pimps up," the traditional values of home (as a gendered and idealized place of comfort, safety, and reassurance) and family (as a nuclear, heterosexual unit). It does so, we suggest, in order to justify and contain forms of relationality and rationality that promise, or threaten, to change the meaning of home and family in an extended, dislocated network of intelligent agents into which the subject is absorbed and translated as object and as data.

It is not that technoscientific processes of objectification and containment are in any sense new. However, the epistemological ordering and reordering of subjects and objects, of humans and machines, brings to the fore the need for an alternative, critical, and interventionist ethics of relationality and becoming that does not shy away from asymmetries of power and blocked possibilities. If AmI helps to instantiate a technology of everyday control by naturalizing and humanizing it, then notions of invisibility and agency it embraces must be denaturalized in response. If AmI encloses and reorders the self as data, sealing us into a cycle of self-reinforcement that pushes us right to the technocapitalist edge, then it is important to continue opening out the question of the self and its constituent, relational others in order to see what kinds of relations are facilitated and prohibited in the process and what consequences various enactments of relationality will have, for "us" and "the world" at large.

Affective Computing and the Intimacy of Technology

As argued by Emile Aarts and coauthors, "Ubiquitous computing environments should exhibit some form of emotion to make them truly ambient intelligent. To this end, the self-adaptive capabilities of the system should be used to detect user moods and react accordingly."[61] Developments in Ambient Intelligence draw on this notion of affective computing, associated with the work of Rosalind Picard,[62] enacting what could be described as a revisionist relationship between AI and ALife.[63] If AI is associated with the sort of disembodied rationality that produced expert but ultimately inflexible and unfriendly systems such as HAL 9000, ALife acts as a corrective to this rationalist computational model by seeking to generate—from the bottom up rather than from the top down, or in a more organic, biological way—an embodied rationality that incorporates affect and emotion. Aldiss's David from his short story "Super-Toys Last All Summer Long" (which provided the basis for Spielberg's 2001 film *A.I.*), was a boy who could love.

Affective computing thus involves embedding computers and computational processes into everyday objects that are familiar to us and that, consequently, seem unthreatening. It is a form of both making technology invisible and giving it a "friendly face," wherein the prior fear of the technological other associated with the early sci-fi era gives way to the supposed intermeshing of human and machine. With ambient computing, the (now touchy-feely) machine is not just a human friend—it is also our double. Significantly, affective computing tends to conflate affect and emotion.[64] Embracing new industry-speak about the self's reinforcement via technology—whereby objects and technological artifacts are being designed to reinforce a sense of self-identity by projecting one's memories, achievements, and so on into the domestic environment—it reiterates and consolidates the entrenched position on the supposed autonomy of humans and machines. According to Suchman, this idea is intrinsically connected with the enchantment and fetishization of apparently stand-alone artifacts that mimic human behaviors, helping as it does to mask the labor involved in the production and operation of such behaviors. For Suchman, crucially, affect is thus a mere effect of multiple agencies. It relies on the considerable amount of (human) labor that goes on behind the scenes of live machinic performance: the actions of talking, smiling, laughing, or even depressed robots.

In the context of AmI, the ordinary human user with no expertise in computer science is thus pressed into affective labor, which is understood as "the corporeal and intellectual aspects of the new forms of production."[65] For Michael Hardt, the term "affective labor" draws on previously separate streams of academic research into "gendered forms of labor that involve the affects in a central way—such as emotional labor, care, kin work, or maternal work" and also on "the increasingly intellectual character of productive practices and the labor market as a whole."[66] Affective labor, Hardt states, enables us to think about the production of affect alongside the production of "code, information, ideas, images, and the like."[67] His claim that this particular form of labor reveals a "new ontology of the human" is not entirely substantiated, although it clearly draws on the kind of anti-Cartesian epistemology—a refusal of the split between mind and body, reason and emotion—that the AmI project itself stands ready to assimilate, not least in response to earlier criticisms of AI. Such an easy assimilation of affect in some recent work in cultural theory draws attention to its limitations as the basis for critical intervention. Not only is affect situated inside the emergent technoscientific order, it is also, at best, positioned as a model of mutuality rather than of what we will call later on in this chapter "true relationality." Consequently, it does very little to trouble the traditional models of causation (how entities affect other entities) or, indeed, ontology (the notion of an entity that is sufficiently autonomous to have effects on others). Affects, Hardt points out, "require us, as the term suggests, to enter the realm of causality," but we would suggest that this does not offer "a complex view of causality" simply by virtue of the fact that "the affects belong simultaneously to both sides of the causal relationship."[68] A truly complex view of causality and a more complex view of affect would not allow us to speak of two sides of a relationship, but would rather point to the relationality, "intra-action"[69] or mutual constitution[70] of the entities involved.[71] As it stands, however, the "I affect you, you affect me" model serves the AmI project perfectly well by enclosing the human subject within an affective cycle of self-reinforcement so strongly promoted by the industry.[72]

Ambient Intelligence operates on the assumption of universal emotional needs. These are drawn from "the top" of Maslow's[73] hierarchy of human needs and defined as "not functional or physiological," but rather as "psychological and cultural."[74] They include "the need to represent roots and heritage, to create a sense of belonging and connectedness and to demonstrate personal identity and achievement."[75] Again, this move toward intimacy—toward emotions and emotional needs—springs in part from a revisionist response to the critiques of earlier forms of technoscience. It also operates according to the logic of advertising, creating needs and desires that it then offers to fulfill. By making these needs and desires as general as possible—that is, by associating them with ideas such as one's roots, belonging, and identity—the creators and marketers of need-fulfilling artifacts ensure that the latter will be hard to resist. After all, who does not experience the "need to make sense of oneself and one's circumstances and to be able to communicate that sense of self to others"?[76] Questions of access and the deepening digital divide are of course elided here, as is the mechanism of what Lyon calls "social sorting," which distinguishes between one category of self and another. Although some histories, memories, and life cycles will undoubtedly be privileged over others, the aim is to encompass everything and everyone—including the technical "outsiders," or illiterates—in an *automated system of machine-machine relationality*. This all-encompassing ambient technological system is then sold to "us" (i.e., those of us who can and will pay for it) on the basis that it will not merely extend our bodies, as McLuhan envisaged, but that it will also extend our very selves into a psychologically and culturally enriched environment.

The trouble with this aspiration, as Greenfield points out, is that "we're just not very good at doing 'smart'" and that, as a culture, "we have so far been unable to craft high-technological artifacts that embody an understanding of the subtlety and richness of everyday life."[77] Nevertheless, AmI aims "to remake the very relations that define our lives, remodeling them on a technical paradigm nobody seems particularly satisfied with."[78] The reason for this lack of satisfaction is that so far, this paradigm has proven to be predominantly functionalist and reductionist. Ambient Intelligence may claim to have transformed human-machine interaction, but we have already seen how, in this context, relationality is reduced to the exchange of data that functions as a commodity. Perhaps the only significant change lies not in the shift toward the technical paradigm but rather in the increasingly personalized nature of the *data that relates*.

The history of mobile and static media privatization is at least six decades long: Andrejevic traces it back to the 1950s. Drawing on Henri Lefebvre's notion of "productive consumption" and Gilles Deleuze's concept of the control society, he shows how 1950s suburbanization offered the route to an increasingly individualized consumption of media and technologies that doubled as monitoring mechanisms and, ultimately, stood for "surveillance-based rationalisation."[79] The suburban home, he writes, "came to serve as the repository for a private set of appliances that displaced or replaced forms of collective consumption: the automobile displaced the trolley and train, the phonograph and radio the concert hall, and the TV set the downtown movie theater. Thus began the trend toward personalization and individuation that eventually yielded the Walkman, the iPod, and the cell phone."[80] The cell, or mobile phone, is thus a personalized version of its predecessor in that it "associates one number with a particular individual, wherever he or she may happen to be."[81] Cell phones can be tracked by GPS technology, which is added to phones as a legal requirement in the United States. But what is "promulgated in the name of security, promises to facilitate the rationalization of consumption" by enabling marketers to locate potential customers in the vicinity of a sale, a favored restaurant chain, and so on.[82] Wireless Internet access allows for on-demand services such as banking and retail but also facilitates "unsolicited, customized marketing appeals" and the kind of cookie-based AI that watches, learns, and remembers not just what any consumer has purchased but also when and where they have done this—and then relates this information "to the behavior of everyone else."[83] Again, relationality in this context stands for the cross-referencing of the potentially vast databases of personal information that we have already started to generate—for the most part passively, unknowingly, and unwillingly.

As Allon and others have recognized, home-based and, increasingly, mobile media networks, "bring about the possibility of surveillant systems fully integrated within consumer landscapes. Within such systems, every act of consumption and transaction can be recorded and processed, entailing the accumulation of an unprecedented degree of personalised information."[84] This "ontology of everyday control" is not new, she suggests, but rather newly naturalized. Not only have the technologies in question become "soft and flexible"—in a way that GOFAI (Good Old-Fashioned Artificial Intelligence) was not—and not only have they "replaced" or, as we would suggest, *repurposed* panoptic discipline with an emphasis on "individual comfort," convenience, and "creative productivity," but such technologies are also in the process of becoming invisible once again.[85]

If panopticism was ultimately based on an *unseen* but internalized mechanism of self-discipline, how might we understand current forms of governmentality that are less mechanistic and more fluid, less industrial, and more informational, less orientated toward the "masses" and more focused on the individual subject? Elements of Deleuze's control society such as flexibility, fluidity, and even friendly technological agency are undoubtedly in play here. Yet we maintain that they do not replace any particular aspects of Foucault's disciplinary society but rather rework them; indeed, they reappropriate the "environments of enclosure" that incorporate and, as we will argue later in this chapter, exceed institutions such as the family and the home.

Context and Chronology: Return to a Closed World?

The relationship between the technosciences of AI, ALife, ubiquitous computing, and AmI is, as we have already suggested, revisionist and continuous rather than principally progressive. Thus each new project revises and reforms the goals of its predecessors while remaining sensitive to both internal and external critique. ALife, for example, partly springs up from the criticism that AI was too top-down, too rationalist, too reductionist, and too Cartesian in its attempt to separate brain from body. As one of us argued before:

> Artificial life is a development of artificial intelligence that is based on a more biological, less psychological approach to the creation of intelligent machines. The aim here is not merely to simulate but to synthesise intelligent life forms—to "grow" them from the bottom-up rather than to persist with arguably failed attempts to programme them from the top-down. The premise of this particular narrative is that in order to understand life (or intelligent life) it is necessary to put it together rather than take it apart, and so artificial life is presented as an anti-reductionist, anti-rationalist discourse in its turn to, or return to, the body.[86]

Similarly, if one of the major criticisms of ALife is that it failed to get its embodied, autonomous, self-organized, and emergent life forms to scale up to anything remotely resembling insect complexity, let alone humanoid complexity, then, in response, AmI is radically scaled down in its recourse to the quotidian. Gary Hall comments on the overinvestment in the notion of everyday life as supposedly "practical, material, social, political"—a tendency he also identifies in some academic writing, where such a concern with the practical, the material and the political is declared to constitute the "*proper* business of cultural studies."[87] This, he suggests, results somewhat paradoxically in a form of *depoliticization* that is part and parcel of many overt political articulations in the study of culture: "This fetishization of politics and its placement in a transcendental position where the last thing that is raised in all this discussion of politics *is the question of politics*, has at the same time often resulted in cultural studies continuing to, at best, downplay and keep within specific limits, and at worst marginalize and even remain blind to, other means, spaces, and resources for politics and for being political."[88]

We might detect similar logic to be at work in the AmI project. By reifying and universalizing the notion of everyday life embedded in (networked, intelligent) technology, it effectively elides the question of both life[89] and technology. This results in the depoliticization of the relationship between the two and the marginalization of the debate about how that relationship might be figured differently. Darin Barney reminds us that this kind of depoliticization is a facet of technicist and especially technocratic discourse[90]—even, or perhaps especially, when that discourse explicitly repudiates any narrow forms of technicism, rationalization, reductionism, and so on. In the context of AmI, we should thus beware of the wolf in sheep's clothing, luring us with the promise of the familiar (touchy-feely technology with a smiley face) while at the same time reinforcing the most technicist and reductionist notions of both technology and the human.

If AmI does not constitute a break from ALife, or ALife from AI, if they are continuous with each other without being successive or equivalent, how do we account for their historical specificity? Paul Edwards situates AI in the context of Cold War America, military cybernetic systems of command and control, and a "closed world" within which "every event was interpreted as part of a titanic struggle between the superpowers."[91] The closed world operated in the name of containment and "global surveillance" by means of a high-tech military.[92] It centered on "human–machine integration," not least through the development of AI software and the kind of "cyborg discourse" that "helped to integrate people into complex technological systems."[93] Edwards argues that cyborg discourse "functioned as the psychological/subjective counterpart of closed-world politics," contributing to the formation of culture and subjectivity in the information age.[94] With their bodies, minds, and selves refigured as information processors, cyborgs "found flexibility, freedom, and even love inside the closed virtual spaces of the information society."[95]

Historically, ALife was the product of the post–Cold War era—a period marked by, among other events, the destruction in 1989 of the Berlin wall that had separated East and West Germany for decades. In this same year, Christopher Langton delivered his seminal paper on ALife. In that paper the discourse of command and control associated with the Cold War era gives way to one of "nudge and cajole."[96] Yet that latter discourse, seemingly more gentle and affective, is still articulated as a means to an end, that end being the emergence of efficient, useful, intelligent machines whose efficiency, usefulness, and intelligence are simply refigured in relation to biological rather than psychological processes. Those processes "tend to be characterized as feminine rather than masculine because they are more distributed, more co-operative, more chaotic, more based on the corporeal than the conceptual and more about growing and nurturing than programming life."[97]

Is Ambient Intelligence, as a direct continuation of ubiquitous computing, a post-9/11 technology? Greenfield certainly recognizes the potential security applications, but David Lyon looks at broader developments in surveillance technology that were "already in progress" prior to 9/11 and that then subsequently accelerated.[98] For him, these technologies are less a direct response to the threat posed by terrorism than they are a manifestation of a particular form of governance. As Lyon puts it, "The desire of several governments to hold on to some semblance of social control, which some felt had been slipping away from them in a globalising world, now found an outlet in 'anti-terrorist' legislation."[99] As well as paving the way for the possibility of video surveillance equipped with facial recognition software, biometric technologies, and smart ID cards, the period after 9/11 also made possible "the convergence and integration of different surveillance systems" used in marketing and policing.[100] Not only are "categories of seduction" created alongside "categories of suspicion," but there is a marked increase in the intrusiveness and opacity of surveillance systems, as "security, we are repeatedly told, requires that information about surveillance be minimised."[101]

If AmI functions more as a manifestation of a mode of governance than as a response to terrorism per se, how might we characterize this mode? For Wendy Brown, it is certainly one in which civil liberties, including the right to privacy, are evaded if not rendered "expendable" in a rationality that justifies all and any means to economic ends.[102] She argues that "while post-9/11 international and domestic policy may have both hastened and highlighted the erosion of liberal democratic institutions and principles, this erosion is not simply the result of a national security strategy."[103] Rather, it is the outcome of a neoliberal rationality that preceded the national and international security crisis. Brown defines neoliberalism as a mode of governance that produces "subjects, forms of citizenship and behavior, and a new organization of the social" according to market criteria and market values.[104] These include profitability, entrepreneurialism, utility and "moral value-neutrality."[105] The model neoliberal citizen, she suggests, is one who "strategizes for her- or himself" among a range of options, not one "who strives with others" to change these options.[106] This subject operates, as Rosi Braidotti points out, a policy of banal "self-interest" and "moral apathy" that mitigates against social justice and "progressive transformation."[107] Such a subject is characterized by Brown as "*homo oeconomicus*":[108] a rational, calculating creature "whose moral authority is measured by their capacity for 'self-care'—the ability to provide for their own needs and service their own ambitions."[109] Here, moral responsibility is conflated with rational action and the effective weighing-up of costs and benefits.[110] Such calculating actions are normalized in a politics that has no outside and no obvious source of opposition, as everything—including the latest technologies of marketing and surveillance—operates on the same principle.

If Brown's outline of neoliberalism may perhaps seem excessively totalizing, and hence immune to resistance, it is not, as we will see, without alternative visions of citizenship, subjectivity, and social formation. Neither is it at odds with Andrejevic's notion of the digital enclosure, which we suggest recalls the closed world and repurposes it for the post-9/11 period, in which security-linked values of marketization are all-encompassing and implemented by increasingly ambient forms of human-machine integration. The digital enclosure, for Andrejevic, is a "monitored space" that "helps rearrange relations of consumption and production," and that constitutes a "forced choice" for consumers and workers alike.[111] Entry into this space, he suggests, may be required by employment contracts and conditions of purchase and may even become the grounds for participation in cultural and social life.

A BBC documentary "Who's Watching You?"[112] confirms this growing trend by recounting, for example, the case of the UK Automobile Association (AA) driver who was obliged by his employers to have GPS technology installed in his car and to account for the timing of every job, down to the minute. He was then sacked when the AA decided that not all of his minutes were adequately accounted for. This driver felt under pressure to watch the clock during his breaks and argued that the monitoring system could not take account of the context of his actions and the time taken up by contingencies such as toilet breaks. His case[113] illustrates the way in which the adage "time is money" can be instituted, especially in private sector industries, where the use of technologies for monitoring purposes remains unregulated. Social networking can also constitute a form of enclosure because the use of Gmail or Facebook, for example, initiates a process of scanning and user profiling that is of value to advertisers who wish to customize their campaigns. Although it is of course possible to avoid using Gmail or Facebook, it is not possible to use those technologies and platforms without accepting the terms and conditions of their use and allowing "every word you type" to be monitored. If advertisers cannot request direct access to personal data (which is converted to categories and profiles), the police authorities can.

What makes every online action and transaction traceable is the unique IP (Internet Protocol) address attached to every computer. This address remains on a database many months after every Google search, and the information it reveals can be traded. Increasingly, the services available online are funded by personal data, and increasingly we, personally, are regarded as legitimate data sources. Every act of consumption produces commodified information that is sold back to us in the form of personalized goods and services. In this invisible cybernetic feedback loop, we constitute ourselves as digital doppelgangers inhabiting an ever-widening enclosure. Andrejevic writes:

Users tend not to recognise just how much information they are relaying to various parts of the enclosure as they read their online newspapers, surf the web, pay with debit cards, or carry their cell phones with them wherever they go. Instead, the devices do the work for them, quietly translating their actions into bits of information that flow upstream along the same channels and pipes that carry data, images, music, and email to consumers.[114]

The cycle of self-reinforcement, that is, the supposed social and emotional enhancement of the self through technology, which is supported by agential devices and machine-machine relationality, is therefore also something of a trap, as it encloses and reorders the self as a marketable (data) object. This reordering happens even as we extend ourselves into networks designed to entertain and serve us, reinforce our identities, and offer "flexibility, freedom, and even love."[115] In what follows, we will argue that it goes beyond what Lyon refers to as "social sorting"[116] and is in the process of producing a subject that is at the same time being retroactively made over and reinforced in a *particular direction*. This oscillation between change and continuity, between a subject in transition and the reinscription of self-identity, lies at the heart of our understanding of mediation as standing both for being-in and becoming-with the world. It is also entirely consistent with technoscientific discourses that make a significant contribution to both opening out the potentialities of life-as-it-could-be and closing them down again in the form of life-as-we-know-it.

Becoming Data@Smart Home

The appeal of Ambient Intelligence as a form of ubiquitous computing is that "of being at home wherever one happens to be."[117] Home, as David Morley explores at length, remains an idealized and gendered category, cleansed of uneven power relations, exploitation, coercion, and incidents of domestic violence. Morley associates home with Heimat, a term that "conflates 'the motherly and the homely'" and "also means 'birthplace,' 'settled,' 'identity,' 'a sense of belonging.'"[118] He rejects criticisms that such ideas and associations are outdated and obsolete, although he recognizes that gender is one persistent instance "of boundary drawing in which households are involved."[119] Morley insists that the linked ideology of housewife and nuclear family still underpins, albeit nostalgically, the late modern household. This argument leads him to interrogate what actually is at stake in this sustained performance of home as haven:

> As Jeffrey Weeks puts it, despite all we know of the actual shortcomings and limitations of life within the traditional nuclear family, we fall back, in the absence of an alternative discourse, on "the security of what we know, or believe, to be secure and stable, a haven . . . where those who feel besieged may find protection." The question is, what price must be paid for this "protection" behind the walls of the purified space of domesticity, and who does the paying?[120]

Along with Allon and Suchman, we have so far suggested that what lies behind the replaying of home as haven—and, indeed, behind the wider processes of reterritorialization and rehumanization that underscore autonomous national and individual identity, both on a biological and technological level—is the attempt to legitimate the increasingly intrusive and intimate forms of monitoring and surveillance. Safe from a globalized, deterritorializing, and hostile exterior world into which our technological agents venture so that we do not have to, we are free to work, play, and shop in our own private and increasingly smart home. "Home-based businesses and telecommuting"[121] are, as Donna Haraway points out, a feature of the newly scaled-down, ambient cybernetic home. Such a home is underpinned, as Lyon argues, by an insidious (because largely invisible) form of social sorting that is automated and effected through algorithmic categories and codes into which the values of the market have been embedded. According to Andrejevic, this systemic architectural form of social discrimination distinguishes between the economically disadvantaged, who are "subjected to surveillance in the form of policing and exclusion," and the relatively wealthy, who are "subject to productive forms of surveillance" and monitored so that they can be offered "individualised goods and services."[122] Both of these groups "do the paying," then, but they do so in different ways. Systemic social sorting is flexible and adaptive. Courtesy of the integration of surveillance mechanisms designed for marketing and policing, it can be tailored to criminals as well as consumers. Lyon relates the case of Mohammed Atta, one of the 9/11 conspirators, who was easily identified after the event by the data trail he left behind: "He could be seen on grainy CCTV film footage entering a motel, paying for fuel at a gas station, picking up supplies in a convenience store, and so on. Not only were there images; his transactional data had been retrieved, showing his online air ticket purchase, his phone calls, his email use."[123]

Such everyday activities have always constituted "intelligence" for the authorities, but they are now available much more quickly and readily. After 9/11, police were able to construct a profile not only of Atta and other individuals involved in the attacks, but also of the new criminal type defined as "Muslim-Arab."[124] This type then became a category in automated systems, for example at airports, which now enact a racialized form of sorting: "For some, the new 'racial profiling' is an extension of the common notion, known all too well by some groups who live in urban areas, that 'driving while black' is likely to attract police attention. . . . Since 9/11, the new category for suspicion is 'flying while Arab.'"[125] Lyon maintains that although racial and other interests are built into the network, they cannot be attributed to the network, or to technology itself. Instead, they are attributed to an assemblage of agents, especially to free-market, socially conservative governance that regards social order not as "a matter of shared values but of smart arrangements that minimize the opportunities for disruption and deviance."[126] Echoing Wendy Brown's sense that Islamic fundamentalism is regarded primarily as an "impediment" and potential disruption to "market rationality and subjectivity,"[127] Lyon states that "if the system works more smoothly, then the fact that it might exclude whole groups of people is not a major concern."[128]

Exclusion is one form of contemporary social sorting that is racial, gendered, and, above all, economic. Given that they are taking place in wealthy, technologized nations, the everyday actions and transactions of the home-based data subject are more likely to be subjected to productive surveillance that classifies and profiles consumers according to market criteria. RFID (radiofrequency identification) tags—one of many existing AmI technologies[129]—promise to be useful in that regard. These are miniature transmitters attached to products, and they contain information about their cost, date of manufacture and sale, and so on.[130] This information can be read and logged on a database. Joseph Turow reports that although RFID tags are currently used to track large consignments of goods, research and development is underway on the potential of tagging individual products in order to track both their use and their users. Ambient environments in the home and elsewhere could be embedded with RFID readers that would continually monitor the behavior of individuals who are wearing or carrying items which have been equipped with RFID tags. Again, relevant information—from brand preferences to gaming habits to shoe fetishes—would be fed back to the network in order to produce customized marketing appeals.

RFID tags are one of the key constituents of smart clothes, the wearable technologies that enable AmI to move out of the home without losing any of the connotations of safety, autonomy, identity and belonging. Wearables expand and enhance the body, rather than the home, as a medium. They help constitute a mobilized interior that is consistent with the car and other forms of what Raymond Williams termed "mobile privatization." The idea behind this project, which stems from nineteenth-century visions of the home, is still of "a shell which you can take with you . . . to places that previous generations could never imagine visiting."[131] The shell is simply shrinking to fit. As it comes to form a "mobile cockpit for digital flaneurs," it crosses a gender boundary, or finally feminizes the modernist masculine flaneur—typically observing others but not being observed himself.[132] Ambient Intelligence both observes and embodies this subject by "storing detailed records of . . . [his] movements, preferences, encounters, purchases, and even vital signs."[133] Like Greenfield, Andrejevic sees in AmI the potential for biometrics and for neural marketing, as our physiological reactions to adverts and entertainment can be inventoried and stored in databases.

Home-based and mobile AmI threatens to realize Bill Gates's vision of a "fully documented life," one characterized by intimate and intensified mediation that is very much a means to an industry- and capital-defined end, even while it is justified in terms of the user's own self-interest, comfort, and convenience.[134] On the one hand, as Andrejevic puts it, we are presented with "the prospect of a customised digital servant that looks after the user." Yet on the other hand, we are offered "clothes that do the bidding not just of their wearers, but of their manufacturers."[135] How will we settle or even preempt this struggle for power? What strategies might be available to users if they are to challenge the presumption that the development of these technologies is inevitable? Are those strategies, as Greenfield suspects, already overdetermined by industry imperatives? Last but not least, what are the technological limitations and political alternatives to this manifestation of neoliberalism?

Everyware Is Nowhere?

If the prospect of intelligent artificial life forms—that is, novel agent technologies that merged AI and ALife—was more "a facet of communication . . . than computation," the manifestation of an inherent *desire for life*, then with the advent of Ambient Intelligence that desire may have been replaced by a demand.[136] Yet demand, as Greenfield suggests, is actually one of the inhibiting factors in the development of "everyware" technology, as "there is barely any awareness on the part of users as to the existence of ubiquitous systems, let alone agreement as to their value or utility."[137] This lack of basic public awareness, let alone demand, signals that, from a manufacturing point of view, *everyware is currently nowhere*. A move toward AmI-driven future is currently being held back by the lack of operating standards covering multiple devices or a concerted program of design and by an adequate infrastructure.[138] Nevertheless, we must account for the performativity of the technoscientific discourse that makes it look not only like we really need AmI but also like it is already happening to us: witness its extension into popular culture (which functions as both stimulus and critique) and the partial realization of an otherwise hypothetical project in the form of portable devices with wireless Internet access, cars and phones with GPS technology, RFID tags, and a host of online monitoring and tracking devices. To the extent that these technologies are already operational, we are faced, as Greenfield suggests, with political and ethical questions that require a response from us—if we are not to accede to an industry vision and its asymmetrical, discriminatory and humanistic, yet ultimately *inhumane*, terms.

In other words, if we do not want "a toaster that talks," or a door that tries to sue us for using it without paying, and if we do not want to triple the time it takes to perform ordinary domestic tasks by having to engage with, program, and convey our preferences to a surreal, irretrievably science-fictional array of animated artifacts with one-track minds, then we had better figure out a way, Greenfield urges, either to opt out or at least have a say about what uses and values they should be embedded with. He insists that

> if you still want to use an "old-fashioned" key to get into your house, and not have to have an RFID tag subcutaneously implanted in the fleshy part of your hand, well, you should be able to do that. If you want to pay cash for your purchases rather than tapping and going, you should be able to do that too. And if you want to stop your networked bathtub or running shoe or car in the middle of executing some sequence, so that you can take over control, there should be nothing to stand in your way.[139]

Drawing on McLuhan's warning that "every extension is [also] an amputation," Greenfield appeals to, and reminds us of, our own judgment and our own, at least partial, residual agency.[140] Similarly, Lyon urges us to resist the anonymity of social sorting, to literally put a face to it, as "the face always resists mere categorization at the same time as it calls data users to try to establish trust... and justice."[141] In asking what it takes to mobilize opposition to surveillance as social sorting, he reminds us that its increase is not just attributable to certain new technological devices: "Rather, the devices are sought because of the increasing number of perceived and actual risks and the desire more completely to manage populations."[142]

What lies behind the faltering, nascent demand for AmI and related technologies is the desire to manage populations according to a narrow market rationality that has fed on and absorbed the security crisis and has grown stronger in the wake of 9/11. Even though this tendency seems unstoppable, Brown suggests that there remains the possibility of "an alternative vision," of a "counterrationality" predicated on the rejection of *homo oeconomicus* "as the norm of the human," as well as "this norm's correlative formations of economy, society, state, or (non)morality."[143] We would argue that recent critical work from within the nexus of feminist technoscience studies, standpoint epistemology, and bioethics may have something to offer in the struggle against the "serious political nihilism" that Brown identifies in the current climate, and that underlies this seemingly innocuous field of AmI. In working on, with and against those command- and control-driven fields that preceded AmI—in taking on cybernetics, AI, and ALife, and in thinking about gendered human-machine, subject-object, self-other relations in ways that could, in Haraway's terms, make a difference—transdisciplinary feminists, philosophers, and technoscience theorists have gone some way toward envisioning what Brown calls "a different figuration of human beings, citizenship, economic life, and the political."[144]

Figuration is a concept most readily associated with Haraway, but it also plays an important role in the work of Braidotti. Understood as ways of shaping a political imaginary, or performing an alternative image of the future, key figurations for those two feminist thinkers include "the cyborg," "nomadic subject," "modest witness," and "companion species." The defining aspect of these and other figurations is what we could call *true relationality*; that is, the way in which they pose a radical challenge to the subject-object dualism that structures Western philosophy—which allows us to believe that humans and machines are separate entities that need to reintegrate and which centers on the sovereignty of the self versus the other. *True relationality challenges the very concept on which neoliberal rationality is based: the (implicitly white, male, middle-class, and heterosexual) self.*[145] True relationality is not an ontological end in itself but should rather be seen as a problem. It is, indeed, what for Bergson would be a "true" problem. "True," as derived from Bergson's understanding of time as duration, does not stand for any absolute metaphysical "truth" but rather means "living" and "concrete": something that can first of all be experienced from amid things, rather than analyzed by a detached observer.[146] This kind of "true relationality" constitutes an insurmountable challenge to philosophy and politics as we know them, undermining their central category of thought and de-essentializing the ontology of the self.

Neoliberalism cannot work without such an ontology of the self, which it aims to repurpose and reinforce but not undermine. As Suchman so clearly illustrates, selfhood understood as autonomous agency is difficult to dislodge in practice, as it bounces back, reinforced, from the relational challenges posed by increasingly animated "intelligent" artifacts and media. Suchman's answer to the repurposing and reenchantment of the self in technoscientific discourses such as AI and AmI comes in the form of a sustained critique, a demystification process that includes shifting our focus from the enchanted agent in question to its context, frame or field of action. For her, Haraway's cyborg was always a little too attention-seeking as an entity, obscuring, as other apparently singular figures do, "the presence of distributed sociomaterialities in more quotidian sites of everyday life."[147] Yet, arguably, the cyborg was never just a singular entity but rather something hybrid, nomadic, in progress. Suchman thus unnecessarily and all too quickly fixes, or "spatializes," what was predominantly a temporal phenomenon. Her ethnographic strategy, centered on quotidian sites of everyday life, is an effective form of disenchantment, even though it takes us into the terrain that the new technoscience is claiming as its own. That, of course, can be readily justified as a good reason to go there—to engage in critical contestation. However, the manner, or mode of the contest should, we suggest, remain much more open than she allows it to be.

What Suchman refers to as "the magic of the effects created" by animated artifacts can be countered by exposing the sleight of hand at work in producing such effects and by revealing the magic as a mere trick.[148] This kind of demystification, an exposition of the practices of AmI and of the industry claims that shape it, has been our ambition in this chapter. Ambient Intelligence is, after all, a particularly nasty trick. Yet on another level—that of talking toasters and treacherous toilets—AmI has a potential to self-demystify or even self-destroy. This potential has been picked up by some of the key figures in science fiction. For example, in *UBIK*, Philip K. Dick already prefigures what we now call ubiquitous computing or Ambient Intelligence. At the start of each chapter, he tells us what UBIK is. It is variously beer, coffee, salad dressing: "safe when taken as directed."[149] It is also a polish ("saves endless scrubbing, glides you right out of the kitchen!"),[150] deodorant, foodstuff, and bra ("supplies firm, relaxing support to bosom all day long when fitted as directed").[151] Finally, it is something resembling God and, as the story unravels, the miracle elixir of a half-life. Such stories are not just amusing: they are also effective forms of critique that counteract one narrative with another. They might well contain a counterrationality, especially where, as in the case of feminist science fiction, the political intervention is explicit.[152] The issue that we are alluding to here is therefore methodological; it is concerned with *how* critical interventions into current states and networks of events have been, and can be, made. In chapter 7 and the book's conclusion, we will outline what we see as the critical effectiveness of interventional forms of creativity that work across residual, institutional divisions between theory and practice, fact and fiction. With this, we do not seek to be prescriptive or to invalidate the more conventional modes of academic critique that have been deployed in this chapter and throughout the book. But here we have also drawn on science fiction as comedy, critique, and a performative co-constitutor of technoscience. If Arthur C. Clarke wrote the critical account of AI that helped to reshape the field, and if Philip K. Dick warned us, it would seem effectively, that *ubiquitous* means *everywhere* and *nowhere*, who will write the alternative story of Ambient Intelligence, and to what effect? The relationship between science and fiction as conventionally conceived (in other words, separate and autonomous) can never be symmetrical—it can be only representational. But what if we cannot really tell them apart? We also have a stake in tackling technoscience in a way, or a mode, that recognizes something feminist philosophers and technoscience scholars in particular have emphasized—namely, that it is not enough to break down dualisms and reorderings such as human/machine, subject/object in theory, as if theory *were not always already a form of practice*. Hall argues that the two are implicated in each other as their mutual conditions of possibility, but also suggests that many so-called theoretical texts "are capable of functioning as singular, active, affective, affirmative, 'practical' events, gestures, and interventions into the here-and-now space of history [and] culture."[153]

The mutual implication of theory and practice makes it impossible for us to conceive of an Ambient Intelligence environment that is populated by servile animated objects and appliances without drawing on the legacy of science fiction in its more skeptical, less celebratory mode, in an attempt to stage something we could call "creative theory."[154] This is also why we have argued that Ambient Intelligence has a significant stake in the play of autonomy and relationality that involves a working out of the question of technological agency. In its industry-led forms, AmI operates as a form of productive containment, closing down on the potentiality and temporality of subjects and retroactively transforming the potentially fluid and metamorphic self into the marketized self, the becoming data-machine. This emergent identity, paradoxically *edged*, in Braidotti's sense, with technology and capital, is a salutary reminder that transformations, cuts, and reorderings occur alongside and as part of the ongoing multilayered technoscientific processes that are in turn part of the broader technocapitalist network of forces. We thus need to remain vigilant in our efforts to repoliticize technoscience as it disperses ever more intimately into our lives—welcomed or otherwise.

Such efforts should involve facing ethical questions that arise with regard to the self and its constituent others—including the question of what we mean by "constituent" and whether what is on the other side of the agential cut (machine, animal, human, alien) is really separate from us at all. For so-called immanentist philosophers such as Bergson and Deleuze, ethics *is* relationality, which involves the recognition of agential connection between emergent agents on the genetic level of life. However, relationality is not progressive or ethical in itself: in AmI, it figures a disconnect, an unethical containment of the self within an automated system that is geared toward the self's marketization. The latter is achieved by means of self-reinforcing affective technological artifacts. If it is not to be a banal form of vital continuism, the unfolding of life and the emergence of agents in and through relationality therefore requires a "cut": a decisive in-cision that is also a de-cision, a meaningful yet temporary resolution within mediation, performed from the midst of it.[155]

5 Sustainability, Self-Preservation, and Self-Mediation

The Banality of Self-Interest

In the previous chapter, we argued that the technologies of Ambient Intelligence and the smart home can be described as predominantly conservative, in the sense that they reinstall and reinforce a traditional, humanist, solipsistic, and antimachinic vision of the self. When the transformation of individual subjects does take place within the remit of those technologies, it is mainly on the level of their repackaging and repurposing *as data*: a process that remains clearly inscribed within a market rationality. In this chapter, we explore the possibilities of a more inventive approach to the self, one that breaks with neoliberal market economics and, specifically, with what Rosi Braidotti terms "the banality of self-interest"—a moral apathy she identifies with "neo-conservative political liberalism in our era," which entails a "hasty and fallacious historical dismissal of social reformism and critical radicalism."[1] Our focus will be instead on *becoming as an ethical practice of emerging-with-difference in the space of relationality*: that is, on the mutual constitution of self and other as and through the process of mediation.

Becoming is a concept frequently employed by so-called process philosophers, of whom Bergson is one. For Bergson in *Creative Evolution*, becoming is conflated with time, movement, creative evolution, and life itself. Elizabeth Grosz describes this set of theoretical maneuvers or even slippages as a form of "opening up" to the outside, "which is at the same time a form of bifurcation or divergence."[2] Becoming offers a way of describing life as an ongoing, open-ended process of differentiation and individuation, a process of creative evolution marked by the "invention of forms ever new."[3] It presumes an element of novelty and indeterminacy, while also positing a kind of perpetual motion in which a reality that denies the division between subject and object, self and other, "makes itself or ... unmakes itself, but is never something made."[4] As well as acknowledging the limitations of becoming as a concept that can perhaps be seen as *too* invested in establishing and maintaining the division between time and space, life and matter, we will focus on the limitations of becoming as a *continual* process of evolution and transformation. Instead, we will concur with Braidotti that becoming must be combined with a degree of self-preservation, or "sustainability."[5] However, we will also explore the extent to which sustainability—as a delicate and perhaps precarious ethical balance between inside and outside, self and other—is exceeded in the process.

It is this idea of a sustainable self that nevertheless does not unquestionably return to any of its prior humanist anchor points that is the focus of our enquiry in this chapter. If processes of intelligent mediation centered on the smart home never fully depart from the humanist idea of the self that can be seen as "banal" in as far as it is presumed to be autonomous and therefore confined to its own sameness, then what can be said of processes of *self*-mediation that are centered on the face, a corporeal dwelling understood not just as an envelope or screen *for* "the self," but rather as a dynamic network of cellular and psychic relations in which the self becomes "itself"? In the processes of corporeal transformation that involve both fragmentation and coming-together in facial surgery, self-mediation is both enacted and writ large—it is literally performed on the body as a medium. Bernadette Wegenstein highlights this dynamic tension between splitting-apart and bringing-back-together that underpins the Western concept of the body when she argues that "the concept of bodily fragmentation, in circulation since the sixteenth century, has, in the twentieth century, been integrated into a holistic body concept—a concept that reveals the history of the body to be, in fact, a history of mediation." Such mediation is, in Wegenstein's terms, "constitutive"; it is a process "for which *both fragmentation and holism are indispensable modes of imagining and configuring the body*."[6] To what extent can it be argued that these dual processes of fragmentation and making-whole end up emphasizing location and self-identity over and above dislocation and differentiation, and what can the consequences of such a valorization be? Last but not least, if cosmetic and reconstructive facial surgery can be understood as the pursuit of the self as an idea and an ideal, then to what extent is this self inevitably relinquished in the process of face transplant surgery—only to be rediscovered as a less idealized, more livable phenomenon in the process?

The (Re)mediated Self

Bolter and Grusin use the example of cosmetic surgery to illustrate their notion of how the self as a concept operates in relation to media and processes of (re)mediation.

Specifically, they look at how it is repurposed through the dynamic interplay between new and old, digital and analog media: "Because we understand media through the ways in which they challenge and reform other media, we understand our mediated selves as reformed versions of earlier mediated selves."[7] Leaving aside for the moment the question of what a mediated self is—a question that of course remains tied to Bolter and Grusin's idea of mediation—the earlier versions of selves they refer to are those associated with the shift from Romanticism to modernism. Romanticism is understood by them in terms of its emphasis on self-expression; modernism is posited as a reflexive move, whereby the sense of reality, authenticity, and presence afforded by self-expression is predicated not on the movement or projection of the self into the world but rather on the reverse movement. Where self-expression is facilitated by visual (digital) media, we retain the option, they suggest, of either passing through the frame or screen into the represented world or of remaining where we are and having that world come to us and surround us. Bolter and Grusin seek to show how such a dual notion of the self is still with us today, as the legacy of Romanticism and modernism. They also inscribe it in their framework of immediacy and hypermediacy, which is indicative for them of the "double logic" of remediation. The remediated self as a reformed version of earlier mediated selves ceases to be a category of thought at this point and becomes naturalized in the oscillation between a subject's relatively stable point of view and his or her shifting networks of affiliations. They write:

> There are two versions of the contemporary mediated self that correspond to the two logics of remediation. When we are faced with media that operate primarily under the logic of transparent immediacy (virtual reality and three-dimensional computer graphics), we see ourselves as a point of view immersed in an apparently seamless visual environment. . . . At the same time, the logic of hypermediacy, expressed in digital multimedia and networked environments, suggests a definition of self whose key quality is not so much "being immersed" as "being interrelated or connected." The hypermediated self is a network of affiliations, which are constantly shifting. It is the self of newsgroups and email, which may sometimes threaten to overwhelm the user by their sheer numbers but do not exactly immerse her.[8]

The authors of *Remediation* insist that the two logics of immediacy and hypermediacy are "alternate" but "complementary" strategies of self-expression, both equally geared toward the goal of achieving a sense of reality, authenticity, and presence.[9] The hypermediated, networked self merely "encompasses and multiplies the self of virtual reality" and is comprised of the "self that is doing the networking" as well as "the various selves that are present on the network."[10] In this account, the self remains an autonomous entity, either immersed or multiplied within a media environment that remains objective for it—an environment that it predates and from which it is epistemologically and ontologically separated. What Bolter and Grusin are presenting here is therefore a representational and spatialized account of the remediated self that interacts with, but *does not intra-act within*, a network of affiliations, and that is transformed only to the extent that it is obliged to multitask. Such a reading seems somewhat limited, as it fails to capture the dynamic constitution of agents and agency in contemporary networked environments in which questions of human-machine relationality continue to be posed. Arguably, there is more at stake here than the delights and difficulties of multitasking. Most problematic perhaps is the emphasis in Bolter and Grusin's account on desire, read as the subject's ability to *choose* different forms of media, to switch from one to the other in the quest for "real or authentic experience"[11]—which in turn reinforces the sense of self. They put it in the following terms:

> The desire for immediacy would appear to be fulfilled by the transparent technologies of straight photography, live television, and three-dimensional, immersive computer graphics. Such transparent technologies, however, cannot satisfy that desire because they do not succeed in fully denying mediation. Each of them ends up defining itself with reference to other technologies, so the viewer never sustains that elusive state in which the objects of representation are felt to be fully present. Our culture tries this frontal assault on the problem of representation with almost every new technology and repeatedly with the familiar technologies. When that strategy fails, a contrary strategy emerges, in which we become fascinated with the act of mediation itself.[12]

It is clear that "mediation" in this instance is set against "presence" and against this nebulous thing referred to as "the real." Mediation is regarded as both a form of representation and a strategy of the self, whereby the self is understood not only in humanist terms but also in neoliberal ones, as a consumer who is free to make a rational choice about the vehicle of its own authentication.[13] This reading of mediation is challenged by more performative accounts of selfhood that scholars in media, communications, and cultural studies are well familiar with, accounts in which the self does not precede or remain independent from processes of mediation but is rather called into being in and through those processes. Indeed, elsewhere in the same volume Bolter and Grusin themselves offer a more performative account of mediation and of the remediated self. They intimate that there is no spatial segregation between subjects, objects, and media and that therefore "there is nothing prior to or outside the act of mediation."[14] Mediation is no longer regarded as a form of representation that impedes access to the real but is rather presented as a performative process that is *constitutive* of the real: "mediation is the remediation of reality,"[15] they say.

Yet at the same time Bolter and Grusin do not sustain this more performative and dynamic reading of mediation itself throughout their argument, even though their central concept of remediation establishes the dynamic interplay between contemporary and historical media. In their discussion of cosmetic surgery, they return to a more representational account of self-mediation, rooted as it is in the idea of the body as a medium and as a vehicle for self-expression. Again, self-expression through cosmetic surgery accords with the double logic of Romanticism and modernism, immediacy and hypermediacy, in their account: "In pursuing the ideal of natural beauty, surgeons must fragment and isolate parts of the patient's body. Although cosmetic surgeons appeal to transparent immediacy, they end up pursuing a strategy of hypermediacy, which calls attention to the process of mediation. What better example could we find of the way in which the desire for immediacy passes into a fascination with the medium?"[16]

The fact that the authors credit this strategy of the self to the surgeon rather than the patient is of note. It functions as a rejoinder to feminist critiques of cosmetic surgery as a gendered instrument of what Foucault termed *biopolitics*: the operation of disciplinary power at the level of biological as well as political life that functions to establish norms and promote cultural ideals—including, in this case, the cosmetic ideal of "natural beauty." Bolter and Grusin's argument is that when this ideal fails to appear, to make itself present in an apparently natural and unmediated way as the outer expression of the subject's true or interior self, then it is pursued—by the surgeon—as a manifestation of the surgery itself. The surgeon is then guilty, very much in line with the myth of Pygmalion,[17] of literally transforming his patient into an object of cultural fantasy—an ideal that she cannot measure up to but must come to represent, courtesy of his skill as an artist, a sculptor. Thus, although the self is remediated through the medium of the body, recalling and rivaling "earlier cultural visions,"[18] it is remediated from the point of view of the surgeon who enacts the sovereignty of the (male) gaze and who occupies the privileged role of the subject in the economy of power-knowledge. From this point of view, cosmetic surgery functions more as a technology of domination—determining "the conduct of individuals" and submitting them "to certain ends or domination"[19]—than as a technology of the self, whereby individuals would undertake a number of procedures on their minds and bodies in order to achieve a state of well-being. Bolter and Grusin therefore ultimately fail to take account of the patient's perspective, presenting her as what Foucault termed a "docile body," an objectified victim of the surgeon's strategy of authentication—that is, his desire for immediacy.

The wider academic debates on cosmetic surgery frequently aim to repair this injustice of perspective and analyze the process from the patient's point of view instead. Many such debates seek to take account of the relation between technologies of domination and technologies of the self, the formation of the state and the formation of the subject (a relation that Foucault referred to as *governmentality*), and the way this relation is played out on both an individual and social level. Governmentality involves complex and multimodal governance of the minds and behaviors of the population—which for Foucault consists of the *subjects of right* who are at the same time *economic subjects*, or actors[20]—via various sociopolitical institutions, a process that is simultaneously restrictive and enabling.[21] It a process in which the modern state and the modern subject are mutually co-constituted and through which they coemerge. Seen from this perspective, cosmetic surgery can be said to be underpinned by a neoliberal drive for enhancement and the accumulation of goods. It becomes an aspect of a wider makeover culture that, in the pursuit of a cultural and economic ideal of perfection, always threatens—or promises—to exceed it.

Distributed Embodiment between the Body, the Image, and the Screen

In an article titled "The Case of the Vanishing Patient? Image and Experience," Mildred Blaxter provides an account of her own experience of being diagnosed with lung cancer. In doing so, she identifies a concern that is currently being voiced in the field of cosmetic surgery, and in medicine more broadly: that "the new technologies of medicine privilege the image over the actual body and its experience, so that the patients themselves may 'vanish' behind the images."[22] According to Blaxter, we are currently being supplanted by "a panoply of screens"[23] that result in us "turning ourselves outside in,"[24] or modeling ourselves on what we see in the image. The idea of a culture saturated with images, and of real bodies and experiences disappearing into a ubiquitous screen, goes back, as Blaxter points out, to the writings of late twentieth-century philosopher Jean Baudrillard. She also argues that this idea is currently being contested within cultural and social studies of medicine, as it does not seem entirely adequate to actual patients' experiences (including her own) of their treatment and diagnosis. Something much more complex seems to be going on in these processes of image transfer in the medical context and it cannot be reduced to a one-way traffic between image and its simulation, between reality and its disappearance. If for Baudrillard the relation between body and image is one of implosion, for Bruno Latour—who offers a much more "networked" and interwoven theory of medical and corporeal transactions—it is one of "translation" and mutual transformation.[25] Similarly, Maud Radstake speaks of "distributed embodiment"[26] as opposed to the disembodiment of the patient in the image. Blaxter presents her own case study as one in which "the patient was not 'vanishing'" but was rather "becoming more clearly defined"[27] through a succession of diagnostic images. It was not the images themselves that left her with feelings of alienation, but rather their translation into records and decisions that seemed somehow to elude both her and any individual doctor's control.[28] Citing Charis Cussins, she identifies and calls for further research into the kind of "ontological choreography" in which images and records "appear to create and control both medical practise and the patient's medical experience."[29]

Blaxter's argument is suggestive of a more complex view of image-centered culture than Baudrillard arguably supplies, one that is less dismissive and more performative with regard to (medical) practice and experience. This more complex and perhaps more problematic view is evocative of Susan Sontag's position when she states, in a somewhat different context, that "to live is to be photographed."[30] Sontag is also referring to the translatability, the mutual transformation and entanglement of embodied lives and photographic images, yet the images she discusses circulate in the sphere of politics rather than medicine and concern not disease but torture and pornography. She argues that the events at Abu Ghraib were informed "by the vast repertory of pornographic images available on the Internet." The events were, moreover, "in part designed to be photographed."[31] What happened at Abu Ghraib was to her a symptom of what is happening in contemporary image-based cultures generally: events and actions, including those performed on our own and others' bodies, are literally informed and ontologically choreographed by images. They are translated from and into images in a way that incorporates—but also goes beyond—representation, toward what we could call "transformative co-constitution."[32]

Although something of this sense of the co-constitution of bodies and images is already conveyed in many current debates on cosmetic surgery—Virginia Blum, for example, speaks of "the cultural goal of becoming photographable"[33]—it is invariably eclipsed by representationalism, a long-standing (but of course not uncontested) belief that theories and accounts of the world provide a credible picture *of* the world, rather than being seen as ways of co-creating this world.[34] It is also enveloped by a pathologized discourse of narcissistic identification in which the cosmetic surgery patient is seen as aspiring to an idealized self-image. Significantly, Blum also highlights the cultural viability of narcissism as an increasingly normalized state when she writes: "Most analysts of narcissistic and borderline disturbances not only acknowledge that these are the most common mental illnesses of our times but also point out that narcissistic personalities in particular may seem entirely congruent with wide-scale social objectives."[35] Within this culture of narcissism,[36] the idealized self-image to which the cosmetic surgery patient is encouraged to aspire is a resolutely youthful one.

Cosmetic Surgery and the Excesses of Makeover Culture

It is worth noting not only that the practice of cosmetic surgery developed from that of reconstructive surgery and the pioneering experiments of Gillies and other surgeons during World War I,[37] but also that it still retains some of the connotations of restoration work. However, the face that is "restored" in cosmetic surgery is an idealized and normalized one; it is an undifferentiated and standardized visage that signifies youth through the renouncement of the idea of time-as-duration, that is, through the absence or erasure of the signs of aging. Indeed, the latter have come to be equivalent with injury, damage, and deformity, as evidenced in references by surgeons to their patients' "drooping, wrinkles, loss of skin lustre and 'middle age spread.'"[38]

In her book *Skintight*, Meredith Jones develops an argument that cosmetic surgery regulates age alongside gender while also promoting a prescriptive, ethnically appropriate aesthetic. The disciplinary power of cosmetic surgery derives from the media as much as the medical institutions, whereby surgery work becomes part of a "makeover culture" in which "the process of *becoming something better* is more important than achieving a static point of completion."[39] Makeover culture is understood by her as a culture of perpetual improvement, in which the body and the home become interchangeable. Having a face-lift is therefore, to cite a surgeon from Blum's account, "like rehabbing a house."[40] In both cases, makeover and maintenance should be done "early and often" in order to avoid "that horrendous megaoperation to rearrange everything,"[41] which rarely has a satisfactory outcome. Many theorists recognize the underlying market rationale that drives the moral imperative to undergo continual makeover, offering carefully priced and packaged products "like a chin implant to go with a rhinoplasty."[42]

Makeover culture successfully connects the sense of autonomy and personal choice with an external injunction to act by reproducing stories of self-improvement. As Blum puts it, "We need movement, travel, stories about going someplace. The trajectory must be from bad or okay to wonderful. . . . We always head into opportunity. Cosmetic surgery stories are inherently future-orientated, are by their very nature about overcoming obstacles through making a change."[43] In a similar vein, Jones argues that whereas the labor involved in self-improvement is elided in the traditional "before and after" images and narratives, such labor is revealed and actually becomes the focus of television shows such as Channel 4's *Ten Years Younger* and ABC's *Extreme Makeover*—shows that are not so much focused on a goal or end point but rather on the sheer effort of self-improvement itself. Makeover shows "are about *processes of becoming*—processes that begin during surgery and then continue through recovery, grooming and further 'personal growth' in everyday life."[44] Jones uses the concept of becoming knowingly but perhaps somewhat inaccurately, in that she does not capture its full creative potential. Instead, for her becoming is a form of governmentality: it is an aspect of neoliberalism that produces enterprising and entrepreneurial subjects who are committed to self-improvement not in and of itself but rather in relation to a set of normative goals oriented toward "the erasure of alterity."[45] Yet something perhaps gets lost in such a unidirectional and instrumental conceptualization of becoming.

For Bergson, becoming remains at odds with both "mechanism" and "finalism." The former implies a linear, deterministic model of cause and effect, coupled with the misguided belief that it is possible to reassemble the whole from its parts, while the latter concerns the pursuit of imaginary goals and end points. As a form of creative evolution, becoming is therefore inevitably indeterminable and open-ended. It proposes a reality that must remain indivisible. However, as Wegenstein suggests, cosmetic surgery in the way it is framed by traditional medical practices is inherently mechanistic, perpetuating a discourse of body fragmentation that can be traced back to the early modern period. She writes: "The process by which the body was objectified and isolated through the practice of anatomy already had begun in the fifteenth and sixteenth centuries, producing what we could call the *scientific fragmentation* of the body."[46] Jones herself demonstrates that cosmetic surgery is also fundamentally invested in the logic of finalism: the pursuit of age, gender, racial, and aesthetic ideals that are most rigorously policed through images of celebrities and a discourse of "makeover misdemeanors."[47]

The pathologization of cosmetic surgery is manifest in notions of addiction and excess that are frequently applied in both public and professional assessment of its clients. Such assessment centers on the alleged sickness of celebrity culture and of "becoming-celebrity." Here, becoming signifies identification and personal aspiration—a kind of finalism that, for Bergson, would be at odds with a more processual definition of the term. The aspiring celebrity may, Blum suggests, suffer from what the analyst Helene Deutsch terms an "as if" personality; one that "never stabilizes."[48] He, or more likely, she—Blum considers women to be more vulnerable to this condition because of the degree to which they are expected to conform to cultural norms and stereotypes which themselves are constantly shifting—lacks a stable, core sense of self, and is therefore unduly influenced by image-ideals that circulate in the external environment. Her personality is only formed through "identifications with others, identifications that keep shifting because there is no core personality discriminating and selecting."[49] The myth of a core or true self is not contested or even explored in such narratives; instead, the "as if" personality is regarded as that which demarcates a norm—a diagnosis which reveals that identity is always already a performance, a trying on and contingent stabilization of external cultural roles. For Blum, the "as if" personality marks the outside, the excess of a sick celebrity culture in which we are all positioned as potential aspirants. As aspirants who identify with, desire and imitate "celebrity bodies," we are all supposedly more and more likely to turn to cosmetic surgery and thus, as Blum suggests, become hostile "to our mirror images."[50] Drawing on Freud's notion that identification is ambivalent and can be an expression of rage as well as desire, she goes on to argue: "In a culture that induces strong identifications with celebrities . . . every now and then such identifications will turn deadly. More typically, the aggressivity experienced toward the celebrity image turns inward and claims our own bodies."[51] If externalized forms of aggression are manifest in media exposés of makeover misdemeanors and celebrities addicted to cosmetic surgery,[52] then internalized forms of aggression, Blum suggests, may be revealed through the spread of body dysmorphic disorder. Just like the "as if" personality disorder, this is an extreme condition, a pathology that is being institutionalized and normalized through the association between cosmetic surgery and celebrity culture, and specifically through the surgeon-patient consensus that there is always a feature of the face or the body that requires "work."

Governmentality, Self-Care, and the Pursuit of Perfection

Victoria Pitts-Taylor defines body dysmorphia as "a mental disorder characterized by a person's obsession about a slight or imagined flaw in her or his appearance to the point of clinically significant distress or dysfunction."[53] Although it has been linked with addition to cosmetic surgery, particularly on the part of actual and aspirant celebrities, Pitts-Taylor's aim is to depersonalize this condition and to recognize how it "operates as a medically constructed boundary for cosmetic surgery culture."[54] As part of cultural rather than just psychiatric discourse, body dysmorphia constructs a boundary between cosmetic illness and "cosmetic wellness," between the inadequate and adequate display of self-maintenance and "self-care."[55] Here, Pitts-Taylor draws on the Foucauldian notion of self-care as a manifestation of how the subject negotiates the norms, in this case of health and beauty, that are external, and that act as forms of discipline, regulation, and reform. Self-care is an aspect of "governmentality" that also exceeds it. It concerns the relation between technologies of the self and technologies of domination.[56] Of particular interest is the emphasis Foucault places on ongoing self-transformation and the pursuit of perfection—which, as we have suggested, has been conflated with the idea of becoming in debates on cosmetic surgery. For Foucault,

"Technologies of the self . . . permit individuals to effect by their own means or with the help of others a certain number of operations on their own bodies and souls, thoughts, conduct, and way of being, so as to transform themselves in order to attain a certain state of happiness, purity, wisdom, perfection, or immortality."[57] Yet the Foucauldian care of the self is not a goal-oriented process that needs to be crowned with some kind of "success." Foucault is very clear about it being a permanent task instead—and a deeply troubling and even irritating one at that—when he says: "The care of oneself is a sort of thorn which must be stuck in men's flesh, driven into their existence, and which is a principle of restlessness and movement, of continuous concern throughout life."[58] Drawing on the Hellenistic and Greco-Roman principle of *epimeleia heautou*—taking care of oneself—he recognizes its forceful imperative to promote the self's *constant* working on itself and its "truth." The open-endedness of this task is what differentiates technologies of the self from "the Californian cult of the self," where you can hope to discover "a true you."[59]

Yet practices of self-care that are evoked in makeover culture very much inscribe themselves in that latter goal-driven, finalist formula, with "governmental" institutions (a Foucault-inflected term that refers to all sorts of sociopolitical institutions that exert coercion and power over the minds and bodies of the population as a whole, as well as over individual bodies and minds) actively promoting that goal. Governmentality therefore functions as a way of imposing constraints on the potentially open-ended promise of the technologies of the self. Consequently, individual and collective makeover practices remain mechanistic and divisive; they are directed toward "thoughts," "conduct," "bodies," and body parts. Such practices of self-care cannot thus be seen as "creative" in the sense of giving rise to truly new bodily forms but represent just the opposite of creativity: they merely reproduce the cultural norms and ideals of perfection that have been transferred or translated from the celebrity image to the aspirant body. Those practices are regularly sanctioned, normalized, and commercialized. They incorporate cosmetic surgery and envelop it within a culture of makeover, maintenance, and what Pitts-Taylor calls "lifestyle medicine."[60]

Typically, our cosmetic self-transformations are therefore both purposeful and purposed; they are driven by the individual pursuit of happiness and perfection and remain steeped in a commercial neoliberal rationality. Wendy Brown argues that neoliberalism "is emerging *as governmentality*"[61] and that it "produces subjects, forms of citizenship and behavior"[62] that are oriented toward self-care as an index of "moral autonomy."[63] Within this framework, individuals are induced to take full responsibility for themselves "in every sphere of life."[64] No matter what individual constraints there are with regard to people's skills, education and opportunity, the consequences of making poor decisions or undertaking irrational actions always fall on the individual. Such constraints thus become depoliticized; they are severed from the broader network of political forces and cultural influences and reduced to the question of individual "choice." Jones analyses how this depoliticization of health care is unwittingly played out in *Extreme Makeover*, an ABC TV series (2002–2007) that drew on communities disenfranchised by minimal state or private health care systems and that would leave individuals to succeed or fail on the basis of their own constrained choices and actions. She also shows how the boundary between success and failure, between good and bad cosmetic surgery, is further policed through the displacement of institutional responsibility onto the individual. The latter is expected to display extraordinary self-determination and desire for self-improvement in order to be deemed worthy of the care, attention, and expense lavished on her by the TV executives.

The criteria for establishing "cosmetic wellness" and deciding on who counts as a "good cosmetic surgery candidate"—good candidates "do it for themselves"[65]—are defined within a disciplinary biopolitical discourse of makeover and maintenance that constitutes one of the emergent forms of neoliberalism. In its current incarnation, neoliberalism, as Brown sees it, has no obvious "outside" and cannot therefore be simply resisted or opposed. Yet the discourse of makeover and maintenance constructs its own outside, its own pathologies, in order to posit and safeguard a norm. Cosmetic wellness is established against cosmetic illness, the good cosmetic surgery candidate against the bad, via a differentiation process with explicit moralizing undertones. In the words of Pitts-Taylor, "Cosmetic surgery that is designed to please someone else is a bad idea; life crises are bad times for cosmetic surgery, and impulsive decisions are bad; there is pain and physical trauma to cosmetic surgery, and thinking otherwise is bad."[66] Within a culture of makeover and maintenance, individuals measure themselves and are themselves measured against these criteria.

Creative Becoming-with-Technology

Although a high number of people seem to conform with its prescriptions, and may consequently be categorized as either good or bad, well or ill, abstainers will paradoxically be identified as good candidates for cosmetic surgery precisely because they have made a rational decision that no matter what their supposed flaws, they do not need surgery to achieve a sense of wellness. Some appear to elude categorization altogether or measure themselves against criteria that provide an alternative to the prevailing culture. The relative opacity of these criteria, and of the alternative cultures to which they belong, comes to the fore in media and academic debates on cosmetic surgery.

Significantly, those debates have themselves played a performative role in constituting such surgical alternatives and in promoting certain "surgery celebrities." Two of the surgery personalities most widely discussed in the academic context are the late American pop icon Michael Jackson and the French performance artist ORLAN.[67] In the context of a disciplinary culture that strictly regulates its members' bodies and lives, their particular relationship to cosmetic surgery can signify excess only as an illness—as something fundamentally outside of the norm. ORLAN's publicized and televised surgery-performances (1990–1993), in which the artist mockingly adopted facial features of various art history icons, turn the pain and trauma involved in cosmetic surgery into a spectacle and are therefore interpreted by some as masochistic, as evidenced in Barbara Rose's description of the French artist's practice as "illustrated psychopathology,"[68] as well as in Fred Botting and Scott Wilson's essay, "Morlan," which discusses ORLAN's work as "consumption taken to unbearable extremes," inhabiting "the space of the inexhaustible plus."[69] "From O-O-O-Orlan to Mmm-more-lan, the excess of femininity comes close to the expenditures of consumer enjoyment," write Botting and Wilson.[70] ORLAN is therefore a "bad" candidate, whose requests for surgery are turned down by the majority of surgeons because her performances go "too far" and are too literal in their attempt to satirize beauty norms.[71] In a similar way, both academic and media accounts of Michael Jackson present him as a "bad" cosmetic surgery candidate and patient because his surgery was, at least in part, designed to "please" his bullying father and/or the greedy media industry and to satisfy an image society that remains obsessed with certain gender, age, and racial ideals.

It is not our intention to condone or even rehearse in too much detail these by now well-known arguments. Rather, the point we want to make is that the excesses supposedly demonstrated by these publicly discussed surgery acts may actually outline an alternative vision of becoming-with-technology—one that is not oriented toward the makeover and maintenance of the self but rather toward the gendered, racial, generational, and technological other. This alternative vision foregrounds connectivity and relationality at the expense of the individual moral autonomy, which is highlighted in the traditional discourse around cosmetic surgery. It entails a more hospitable encounter with alterity, and thereby takes some steps toward enacting a "bioethics that can counteract the normativity of the biopolitical regime."[72] If this alternative vision and its attendant bioethics is arguably foreclosed in makeover television programs,[73] it is at best only partially signaled by figures such as Jackson and ORLAN. Both Jackson's and ORLAN's opening up to the outside or the other is never fully open-ended, and may even be interpreted as a form of identification, albeit with the (im)perfect amalgam of racial, gender, and generational types, or with a hybrid or even transgressive ideal such as the cyborg. In as far as they "perform an ethics of welcome," Jackson and ORLAN simultaneously "bear the strain of hospitality."[74] For Jones, they are the martyrs of makeover culture, "suffering for the sake of beliefs, ideals and for others."[75] Although they embody a notion of the self that is excessive, open to alterity and therefore entailing a corporeal and symbolic provocation, they also reveal the difficulty of a long-term sustainability of any such notion. In the current ecological discourse that underpins many contemporary political economies, sustainability is defined as a long-term maintenance of environmental, social, and economic well-being; it is an ability to meet society's present and future *needs*. However, for us here sustainability stands for more than a systemic capacity for survival *in* the world: it can rather be understood as a form of becoming-with but also becoming-different-from our environment.

Transplantation, Metamorphosis, and the Third Face

Let us now take a look at another practice of corporeal transformation that shares many of the material-discursive aspects with cosmetic surgery: face transplantation. This innovative radical procedure has been positioned as an alternative to, and an advance in, reconstructive surgery. Since the first case in 2005, it has been used to treat severe facial disfigurement—even though research in this field emphasizes that the severity of the disfigurement is not a direct indicator of the "need" for a face transplant and that the disfigurement must be considered along with psychological and social factors.[76] Face transplants have been performed when reconstructive surgery such as the use of skin grafts has not been, or could never have been, successful in restoring functions such as the use of the mouth. It has also been used to achieve a more "satisfactory" appearance. Surgeons in France, China, the United States, and Spain have carried out partial and full face transplants. The procedure, as outlined by the Royal College of Surgeons of England, involves attaching some or all of the donor's soft tissue (skin, fat, facial muscles, veins, arteries, and nerves) to the patient's underlying bone structure.[77] The result is what can be termed a "third" or "hybrid" face. The patient does not look like they used to (that is, "like themselves") and neither, contrary to what is suggested in the 1997 action movie *Face/Off* (in which an FBI agent and a terrorist swap faces and assume each other's identities),[78] do they look, or indeed behave, like the donor.

Psychologist Alex Clarke and surgeon Peter Butler are the principal figures in the UK face transplant team. They have suggested that due to the psychological, social, and ethical complexities of face transplant surgery, its advantages and disadvantages should be debated in the public arena. At the same time, they maintain that it is difficult to conduct such a debate because of the science-fictional connotations of face transplantation and the nature of media interest and news reporting with regard to it. Their concerns center on the misinformation and sensationalism that surrounds the idea of identity exchange—an idea that, as they point out, circulated "well in advance of any description of the range of surgical techniques envisaged."[79] Indeed, far from eliding the significance of face transplant surgery, the prior circulation and enactment of ideas in popular culture has played an active role in the development of face transplant policy and procedure. The power of the Frankenstein myth and the prevalence of public misconceptions about the meaning of face transplant surgery are among the reasons cited by the Royal College of Surgeons for preventing surgery from proceeding prior to 2006. The introduction to the original Working Party report published in 2003 suggests that face transplant surgery had always been a highly mediated event. Media speculation about face transplant surgery heightened in the United Kingdom in 2002 in response to a paper by Peter Butler on limb transplants in animals. It "extended to trying to identify the possible recipient of the world's first face transplant" and led the charity Changing Faces to ask for a moratorium on further media coverage.[80] Unable to grant this moratorium, the Royal College nevertheless "shared the concerns" expressed by the charity and incorporated the issue of privacy and media intrusion into its report.[81] The report concluded that the procedure was still too experimental and that the risks associated with rejection were too hard to predict. It was therefore deemed "unwise to proceed" until further research had been done, a state of events that amounted to a "much more incremental approach than some of the current hype surrounding it [face transplant surgery] has suggested."[82]

Clarke and Butler subsequently contributed to further research on organ transplantation, immunosuppression and the role of pre- and posttreatment compliance with drug and other self-care regimes. Their work addresses the myth of altered identity that used to be "the most emotive barrier" to face transplantation.[83] Although recognizing that the face "as the central organ of communication, the focus of sexual attractiveness and the means of immediate recognition by others" has a deeply symbolic as well as functional significance, they also maintain that what is at stake in face transplant surgery is not altered identity per se but rather a process of "repeated alteration in appearance."[84] The recipient's face changes, but not into the appearance of someone else (the donor). Eventually, over a period of years, the new appearance stabilizes and, it is hoped, becomes concordant with the subject's sense of identity. Interestingly, Clarke and Butler introduce a temporal and transformative dimension to the understanding of identity in face transplant surgery. The recipient, they suggest, does not suddenly or immediately look like the donor. In anticipation of the actual event of a face transplant, they argue that what will occur is not identity exchange as such but rather *a process of transformation in appearance and identity*, a process that leads to the emergence of a "'third' face with an identity of its own."[85] This narrative of process and metamorphosis has subsequently informed Clarke and Butler's patient selection procedure as well as the reporting of face transplant surgery worldwide.[86] Using a combination of laser scanning and computer composite imagery, Clarke and Butler modeled the emergence of a third face from two composites: Butler's face superimposed on Clarke's underlying craniofacial skeleton, and vice versa. Not least through the citation and modification of this research in the press,[87] the two surgeons were able to make a successful intervention into "the *Face/Off* scenario," while also addressing in more detail some other aspects of UK policy.[88]

Cooperation versus Noncompliance

Central to the concerns expressed by the Royal College of Surgeons was the need for more research into the psychological and clinical risks associated with transplant rejection and the establishment of a mechanism for evaluating risk versus benefit as far as this form of surgery was concerned. Despite their critique of this "risk-averse approach," modified in a follow-up report that was published in 2006,[89] Clarke and Butler have considered risk-benefit ratios and related them to the problem of measuring the patient's quality of life.[90] It is widely acknowledged, even in quantitative studies, that the assessment of risks and benefits in this context cannot ever be objective. For example, Barker and coauthors argue that "decisions of risk must include input from those taking the risk, because . . . perception of risk and the amount of risk individuals are willing to accept . . . differ widely."[91] Due to the perceptual nature of risk, "it is difficult if not impossible to ascribe an absolute value to the risk/benefit ratio of various transplant procedures."[92] Similarly, Clarke and Butler maintain that "quality of life is a concept that is both difficult to define and even more difficult to measure," given the different values individuals place on various activities, and their sense of loss when these activities are prevented due to illness or injury.[93] At stake are activities as diverse as education, work, sport, socializing, and sex. If there is a more short-term emphasis on saving lives through transplant surgery in the accompanying medical and media discourse, there is also a foregrounding of a longer-term goal of restoring quality of life along with, in this case, facial function and appearance. However, quality of life "cannot be . . . inferred from objective rating of facial appearance and must be systematically explored and measured from the patient's perspective."[94]

Without ascribing an absolute value to the assessment of risk versus benefit or attempting to objectify quality of life issues, Clarke and Butler nevertheless seek a systematic exploration and measurement of both from the patient's point of view. This approach has enabled them to establish a framework for psychological assessment and a procedure by means of which they have been able to select suitable candidates for face transplant surgery in the United Kingdom.[95] The procedure they have established is broadly governmental, to use Foucault's term, and combines institutional and disciplinary measures such as clinical histories and psychometrics with practices of self-care, self-management, and self-responsibility. On the basis of the established procedures for evaluating patients for cosmetic surgery, the ideal face transplant patient is, paradoxically, most likely to be the one for whom surgery is not considered necessary to their recovery and sense of wellness. Indeed, one of the goals of the procedure that Clarke and Butler implement is to deselect patients for face transplant surgery or, failing that, to work with them on managing their expectations, attitudes, understanding, support structures, and compliance.[96] They write: "The extent to which people successfully carry out those behaviors recommended by health professionals has been variously termed compliance, adherence or concordance with treatment advice, these terms reflecting a gradual conceptual shift to an equal relationship or consensus between the patient and the team who are managing their condition."[97]

The problem of noncompliance with long-term immunosuppression was highlighted in the second Royal College of Surgeons report. Noncompliance supposedly caused graft rejection in the case of the first hand transplant and is thought to have led to the death of a face transplant patient in China. In response to this problem, Clarke and Butler have devised a technique for selection that is based on an extensive series of interviews, by means of which "patients are assessed with regard to knowledge and beliefs, motivation and self-efficacy, and coping skills."[98] Coping skills include "help with managing media interest and potential intrusion."[99] Clarke and Butler stress that "the option of doing nothing is fully explored."[100] The patient is given information regarding the risk of graft rejection and that information is then elicited back from the patient:

- What are the chances of losing the graft at the time of the operation?
- "Well it's 50:50."
- Are you sure?
- "Well actually it's 4 in 100—but the way I'm thinking, it will either happen or not, so it's 50:50 and I need to think regarding what I do if it fails."[101]

Finally, the screening procedure produces a plan to minimize noncompliance and maximize self-care, with the plan itself being "based on individual need."[102]

Transplant surgery raises significant questions about change and the extent to which it can be modeled, measured, and maintained. Face transplant patients are presented with "a heightened sense of change" that the UK team plans to manage through models of postoperative care covering periods of two weeks and one year, during which the process of psychological integration with the donor graft is generally expected to take place.[103] However, these models are indicative rather than prescriptive, in part because the procedure has not yet taken place in the United Kingdom, in part because the long-term prognosis for face transplants is still not known, but also because the notion of change that implies integration remains both dynamic and highly contingent in the medical literature on face transplantation. First, the process of integrating the donor graft and subsequent new appearance with the patient's identity adds a temporal and transformative dimension to an already "dynamic phenomenon" understood in terms of body awareness and body image.[104] Second, it is contingent on achieving and sustaining a complex balance of psychological, clinical, and social factors over a duration of a lifetime (certainly more than one year). In other words, change in this context significantly exceeds the plans to determine and manage it. The fact that this inevitable deferral of assessment is, to an extent, recognized in those plans indicates the limitations of the Foucault-inspired theory of governmentality typically used in media, communications, and cultural studies accounts of body makeover and cosmetic surgery to describe those plans adequately.

Indeed, the notion of governmentality has been criticized for outlining a rigid, structural model of government that is unable to take account of change.[105] In their review and defense of governmentality as a productive concept (in a double sense of this term), Rose et al. arguably only reinforce the paradox that a theory predicated on a critique of structure, with all its inherent dualisms (dominance/resistance, top-down/bottom-up, freedom/determinism), takes that structure as its premise and is ultimately confined by it. Governmentality thus ends up setting up, as if in advance, its foundational structures such as the subject, the state, the individual, power, rights, and so on, in order to then proceed to show how these very structures are constituted, modulated, and productively opened up by governmental processes that are enacted on both individuals and populations. It is therefore unable to see any *radical* and *creative* change to those structures, working as it does on keeping the structural elements of the sociopolitical order in place, even if does allow for some transformation within them.

Taking Account of Change

Although the procedure for screening prospective face transplant candidates in the United Kingdom is broadly governmental, it nevertheless exceeds governmentality in as far as it pays serious attention to and tries to take account of change and thereby recognizes its own limitations as a practice of modeling, measurement, and preservation. Brill and colleagues argue that face transplant surgery will "contribute to our understanding of the development and adaptation of body image."[106] Body image for them is a "construct" that is composed of an externalized, social, or objective appearance and a more internalized view of the self.[107] It is interesting to compare this psychological view of body image with contemporary philosophical approaches that also distinguish between the body as an object, image, and representation on the one hand and the body as a more affective and fluid, ambivalent, and incoherent sense of self that is essentially "without an image" on the other.[108] Following Bergson, Mike Featherstone regards the body as a center of "indetermination," always extending beyond itself and not constituted as a "stable pre-formed" entity or as a platform for identification.[109] Significantly, this sense of the body as movement is incorporated into the medical literature on face transplant surgery, something that may enable both the discourse and the practice to make a contribution to psychological studies of body image. Not only is body image revealed to be always already dynamic and subject to change from infancy onwards; it is also combined in this context with integration as a temporal and transformative process that stabilizes contingently but that never ceases. The coming together of a new appearance and a new identity is presented as a goal in face transplant surgery, but it is never an end point. It is assigned an approximate duration (one year) but is understood to exceed it[110] and to be dependent on achieving a balance between psychosocial and clinical factors that are themselves subject to change. Even here, face transplant surgery resembles other forms of reconstructive surgery, with one key difference—namely, that the process of integration incorporates a relation between donor and recipient in which the psychosocial, if not the clinical, stakes are significantly higher than in other forms of transplant surgery and in which questions of ethics and responsibility come to the fore more prominently. As Brill and coauthors put it, "Clearly facial transplantation represents a completely new opportunity to study models of body image. Not only is the individual adapting to change as in other reconstructive procedures such as skin grafting, but the assimilation of change must incorporate the idea that, however minimal, the body image is in some ways derived from that of another individual."[111]

To sum up our argument so far, the procedure for the selection of face transplant patients in the United Kingdom and the plans for managing change after surgery are typically described in medical literature in governmental terms. Yet "becoming" as a form of open-ended transformation—which should not be conflated with what Foucault means by "technologies of the self" that enable a form of governmentality, and that are premised on the individual's capacity for self-control—is perhaps a more adequate term for understanding face transplantation, and particularly the process of donor/recipient integration. For Braidotti, "becoming has to do with emptying out the self, opening it out to possible encounters with the 'outside.'" It is predicated on the ethical ideal of what we have called in chapter 4 "true relationality," whereby the subject seeks the boundary between self and other(s), balancing their responsibility toward both and sustaining openness alongside the need for self-preservation. Occupying the "borderlines, or lines of demarcation, between my and other external bodies"[112] is not, Braidotti acknowledges, always an ethical choice, even if it is an ethical act. It is of course important to consider that for the face transplant subject, these lines or edges are initially—and potentially in the future—associated with trauma. As a result of illness or injury, the patients are confronted with otherness in the form of an altered appearance. The subsequent goal as defined in the medical discourse is to achieve sustainability by integrating the subject's self-identity with their altered appearance over and beyond a specified and managed period of time. The process of integration is transformative in that it gives rise to new bodily forms. Specifically, face transplant surgery allows for the emergence of what the surgeons term a third or hybrid face, with an identity of its own. It also involves experimentation with the possibilities of a hybrid immune system.

The Induction of Tolerance

It has to be acknowledged that immunosuppressant drugs used in transplant surgery have harmful side effects: they are associated with toxicity, infection, and cancer.[113] In order to circumvent these side effects and prevent incidents of chronic rejection that operate in two directions between graft and host and that increase over time, transplant research has included attempts to induce tolerance.[114] Tolerance of the donor graft or organ can be induced clinically by means of bone marrow transplants. As Hettiaratchy and colleagues explain, "Such transplants lead to a situation known as hematopoietic chimerism, where the recipient has hematopoietic cells, including cells of the immune system, of both its own and donor origin. This leads to a fundamental *reprogramming of the recipient's immune system so that it recognizes and accepts the donor tissue as its own.*"[115] Clarke and Butler report that some success has been achieved in inducing tolerance and avoiding the use of immunosuppressant drugs in kidney transplants. However, the procedure involved using high doses of radiation to destroy the recipient's bone marrow prior to the infusion of the donor's marrow. Alternative approaches that do not require such high doses of radiation are more experimental and involve depleting the recipient's T-cells and then injecting donor stem cells into the host bone marrow and thymus. "The thymus is where T-cells are educated and where self-tolerance is determined. The presence of donor cells in the thymus leads to the development of tolerance to these donor cells and hence donor allograft acceptance."[116] To date, this more experimental approach has not replaced the use of immunosuppressant drugs in face transplant surgery, but in some cases at least it has been used in addition to them.[117] This approach is likely to be superseded by tissue culturing, as the grafted tissue will still have some morbidity, which is why the use of the patient's own "autologous" tissue "will always be preferable."[118] It might be argued that the idea of inducing tolerance resonates most strongly in psychological, social, and ethical contexts, although for the time being it also remains highly relevant in a clinical situation. What makes the process of integration so open-ended and contingent is the need to establish and maintain "tolerance" in each of these contexts simultaneously.

The first hand transplant, performed in 1998, serves as a good illustration of the complexity of those relations and articulations of tolerance. A report coauthored by one of the surgeons subsequently involved in the first face transplant[119] in 2005 states that the hand graft was rejected and removed after twenty-nine months because "the patient did not comply with the immunosuppressive treatment."[120] However, it is likely that poor functioning and weak sensation helped to produce a prior psychological rejection of the donor tissue in this and other early hand transplant cases.[121] Certainly, the more recent research has been concerned with the role of sensation in successful integration, focusing more on the psychological and psychoanalytic factors involved in hand transplants.[122] The psychological and social aspects of integration remain inseparable, as the patient's ability to tolerate the graft, particularly in relation to face transplant surgery, depends on the (quite often prior) acceptance on the part of family, friends, and the wider community. It is interesting to consider the extent to which the United Kingdom displays a high level of potential tolerance to face transplants because of the experimental facial surgeries pioneered and performed during World War I and the way in which the injured soldiers were taken in by the local community.[123] The psychosocial aspects of face transplants are the principal focus of postoperative management plans. Early cases have shown that they, in turn, carry with them a strong ethical dimension.[124] Here, ethics does not refer to the calculation of risk and benefit or to the objective assessment of quality of life—both of which are in fact deemed to be effectively impossible—but rather to the openness on the part of the face transplant subject to the outside, that is, his or her willingness to recognize the donor/the other as a fully integrated part of themselves. For Isabelle Dinoire, who underwent the first face transplant, the process of integration involved in no small part an ethical experience. "She found herself telling her daughter 'her nose was itching' and then saying, 'That's nonsense. It's not *my* nose. I have a nose that is itching.'"[125] The process that takes Dinoire from "my nose" to "not my nose" to "I have a nose" is neither linear nor progressive. Ethical tolerance as a hospitable opening to alterity can always be interrupted by clinical rejection, which may be episodic or permanent. The concept of the edge, or borderline, therefore has a dual meaning for face transplant subjects. One such meaning involves trauma, an edge from which the subject must retreat. The other meaning entails healing—which is a complex relational process with no guarantee of success. Both meanings are contained in the scar,[126] the literal edge that marks a face, the cut that is at once surgical and agential—as it shapes up a new face but also carves out a new agent of ethics who has to give an account of, and take responsibility for, her coemergence from the corporeal-medical-psychic network she finds herself in.

The face transplant subject enables us to rethink some of our notions regarding the embodied subject as a process rather than an entity, as co-constituted rather than autonomous, and last but not least, as understood in terms of becoming rather than being. Such a subject shows us that perhaps thinking in terms of such preferential concepts is itself too dialectical, abstract, and idealized. If becoming signals desire, an instinctual push toward life, then the face transplant subject also conveys necessity, coupled with a pragmatic recognition of it. If becoming stands for affirmation, change, and life itself, then the face transplant subject is also confronted with mourning and the loss of a former appearance and identity in the process of its "becoming-other." The face transplant subject therefore demands a third term by virtue of being endowed with "a third face." Also, having a third face is not per se a sign of ethical virtue. The face transplant subject's tolerance for her relationality with the donor is, in part, induced—chemically, immunologically—and is always open to the possibility of rejection. Cuts, as Karen Barad argues, cut "things" and people, "together and apart."[127]

Beyond Sustainability

Although face transplant surgery reconstructs or restores facial function and a culturally acceptable appearance that would not otherwise be possible, it does not restore the subject's prior sense of self. Rather, the goal of face transplantation is to create, through a process of integration with the donor graft, a new self-identity that is internally and externally governed and that remains open-ended, simultaneously stabilized and metamorphic. Arguably, face transplant surgery forms a continuum with cosmetic surgery insofar as it instantiates self-identity and governmentality, but it also becomes discontinuous with cosmetic surgery to the extent that it relinquishes self-identity and exceeds governmentality. The concept of the self in face transplantation does not take the form of an ideal, located either in the past or in the future. Although face transplantation has much in common with other forms of cosmetic and reconstructive surgery, its temporality is nevertheless different. The self in this particular type of body modification is oriented toward stability and autonomy, while undergoing a simultaneous opening to process, relationality, and contingency.

It goes without saying that face transplant surgery does not exist in isolation from other medical, social, and political practices, which is why it does not remain untouched by the culture of makeover and maintenance predicated on individualism and the moral duty of self-care. Indeed, the face transplant patient is obliged to make rational and informed choices about his or her treatment and, like the cosmetic surgery patient, is compelled to act out of self-interest rather than the desire to please or help others. At the same time, he or she is confronted with the prospects of tolerance and rejection that are oriented outward, toward the other, and that are simultaneously psychological, social, clinical, and ethical. It is our contention that the complex, hybrid and precarious nature of the boundary between the self and the donor in face transplant surgery does not fully support the notion of sustainability drawn from developmental and ecological discourse, whereby sustainability, after Braidotti, stands for "a regrounding of the subject in a materially embedded sense of responsibility and ethical accountability for the environments he or she inhabits," promising "the possibility of a future, of duration or continuity."[128] Although the sustainability of life, or what we could describe as its *good* (that is, prudent) rather than *mere* survival, is posited as a value here, Braidotti is fully aware of life's destructive force: "all life is a process of breaking down," she repeats, after Deleuze. Transplanted organs—including, possibly, the skin of a face—are subject to episodes of chronic rejection that increase with time and that lead "to a gradual attrition in organ function."[129] Immunosuppressant drugs carry severe side effects that also accumulate over time, and experiments in hematopoietic chimerism are purely supplementary in this context. Autologous tissue culturing is likely to supersede these experiments, and although it would clearly be preferable from the patient's point of view, it would also obviate the ethical challenge on which the notion of sustainability is based.

Braidotti has been criticized for placing too onerous a burden on her ethical subject who is urged toward the edge of herself but not *over* this edge and who is expected to possess or acquire a very fine sense of balance in order to ensure her self-preservation. That the face transplant subject's balancing act is part managed and part induced testifies to the "dizzying" task she takes on out of a combination of choice and necessity.[130] The face transplant subject literally acts *out of* self-interest. She takes on a burden of responsibility to the donor and seeks to endure clinically, psychologically, socially, and ethically. She accepts the possibilities of both tolerance and rejection, working as she does through what Braidotti terms "the concept of the crack, the visible or invisible mark of unsustainability,"[131] toward a possible future relationality whose outlines and outcomes are by no means guaranteed.

6 Face-to-Facebook, or the Ethics of Mediation

From Media Ethics to an Ethics of Mediation

To recognize that a technology or a medium has some degree of agency is not to assign autonomy to it and thus simultaneously abdicate "our human" responsibility. However, who or what counts as "the human" is undergoing a significant transformation in the new media context, as we have argued throughout the book, which is why the question of agency and responsibility can no longer be perceived, unproblematically, as something that a skin-bound human entity has, or is capable of exercising. Across the increasingly dynamic boundary of media production and consumption, there is therefore a need to rearticulate what has become known as "media ethics" in terms of "an ethics of mediation." This ethics—in line with our expanded understanding of mediation as a vital process, a way of being and becoming in the technological world, or a way of emerging through time—can also be dubbed "an ethics of lifeness." This chapter will thus pursue the ethical implications of the ultimate instability and transience of the mediated subject. It will pose the following questions: what moral frameworks become available within the context of ongoing dynamic mediation, and who does ethical responsibility concern if "we" are all "media"? What is entailed in the recognition that "nobody and no particle of matter is independent and self-propelled, in nature as in the social"?[1]

The pertinence of ethical questions to issues of media and technology seems self-evident these days, in both an academic and a wider public context, given the ubiquity of moral quandaries and moral calls to action with regard to issues as diverse as genetic modification of humans, animals, and plants; Internet privacy; the independence of media from financial and political influences (or the lack of it); and journalistic practice. It could even be argued that ethical issues concerning media and technology are so complex and all-encompassing that attempting to devise a singular ethical framework that would cover them all would be a task doomed to fail at its very inception, and that there would be nothing on an ontological level that would allow for a rigid and rigorous differentiation between "media ethics" and what Western philosophical tradition has understood as simply "ethics": an enquiry into the formation of moral values and forms of conduct in a given culture.

This may explain why what has become known as "media ethics" in communication studies has set itself a more modest but also more manageable task of dealing with specific issues concerning traditional broadcast media, and now increasingly "new" media: issues such as journalism ethos; questions of truth, manipulation, and representation; and the relationship between media content and the law. Positioned in this way, media ethics is a form of "applied ethics," whereby a previously worked out moral theory, or set of rules, principles and values, is applied to specific "cases"—which are then judged according to the rules already in place.[2] With its origins in and focus on journalistic practice, media ethics has often taken the form of a "code of conduct" (of a kind that also exists for physicians or business people), more often than not an idealistic horizon against which "real-life" actions and misdemeanors are to be judged. Indeed, the problem of media ethics very often becomes synonymous with the question, as Matthew Kieran puts it, of "what is it that journalists, morally speaking, should do?"[3] At the same time, there is a certain urgency within media ethics, in the sense that it is frequently being written up and mobilized "in the white heat of the moment"[4] to deal with ongoing developments in the global and local mediascape, and with the "media conduct" that accompanies them. Its standard reliance on predefined moral frameworks and positions—which are being negotiated with regard to specific media cases but do not themselves undergo much of a transformation—may be one way of managing the continued novelty and urgency of media issues.

It is not our intention here to mount a critique of media ethics, especially given that we recognize the strategic need for the institutional, procedural, and legislative interventions it is often capable of making. Indeed, its specific focus on predefined issues is what actually makes such "media ethics" workable, although on a philosophical level what its proponents often attempt to accomplish should perhaps rather be situated under the rubric of "policy" or "regulation." Our aim is to do something different, however: we want to outline a new framework for thinking about ethics with regard to media technological issues, where the agency of ethical subjects and objects is still to be determined, and where there are no predefined fixed values to be applied or drawn on. We seek to do this in order to capture the dynamic coevolution of ethical subjects with the world, a process that we have described with the term "mediation" throughout this volume. Our "ethics of mediation" will thus be nonnormative, in the sense that it will not resort to predefined values or truths in advance (such as democracy, freedom, transparency, dignity, and so on), nor will it posit a moral core from which an ethical judgment can be issued (Christian soul naturally endowed with the sense of good and evil, Kantian reason). Yet it will not be relativistic, either. Inspired by the Derridean horizon of justice,[5] which provides a background for such attempts to think and act ethically without any supreme external or internal guarantees with regard to what is absolutely "right," this ethics of mediation will respond to singular situations and events that will demand judgments and decision from us and from the midst of the mediation process itself—of which we are part.

The difference between what in the Anglo-speaking academic context has become known as "media ethics" and our "ethics of mediation" pivots not just around the different ethical positions they respectively occupy (normative vs. nonnormative, applied vs. always singularly recast) but also around the notion of "media" that underpins both ethical frameworks. As stated throughout this volume, the notion of mediation we are using here positions media not just as a series of objects (computer, iPad) or broadcasting practices (TV, radio, the Internet), but first of all as dynamic processes of emergence in time, and of our coemergence *with* media. This co-constitutive aspect of mediation opens it to a different ethical framework: that of intra-action and mutual becoming. The stabilization of media out of mediation, one that involves making cuts to the duration of things, as argued in chapter 3, calls for a two-level reflection on both the processes of media (co)constitution and on the responsibility that the cuts of and within mediation envelop us in.

Demediation

To illustrate what we mean by this and how this ethics of mediation could possibly be put to work, let us focus on one of the most ubiquitous media environments of the last few years: the social networking site Facebook. The ethical problems that most frequently get raised with regard to Facebook predominantly concern individual human behavior: that of its founder, financial backers, and users, respectively, with other media (film, journalism, blogs) often mobilized to set up and arbitrate over the moral debate. And thus in David Fincher's 2010 movie *The Social Network*, Facebook's founder, Mark Zuckerberg, is supposedly portrayed, in the words of *National Post* journalist Peter Foster, as a "morally-challenged geek."[6] "According to *The Social Network*, Mr. Zuckerberg was a misfit with zero people skills, a spinning moral compass, and a pathological sense of self-righteousness when he founded the company,"[7] writes Foster. The layering and interweaving of media forms and narratives in providing an analysis of the moral dimension of Facebook is interesting; even more interesting is the fact that this multilayered process of mediation involved in the production of what we might call the ongoing, multisite "event of Facebook" gets erased in such narratives. What we are thus ultimately presented with is the question of an individual's morals and the way they are tested, maintained, or compromised in the Facebook environment. The singularity of the moral agents that join the "Facebook event" remains untroubled, as do the very same moral dilemmas we know from the pre-Facebook world: honesty versus duplicity; popularity versus success; sharing versus accumulating. This line of thinking is also evident in the majority of the essays gathered in D. E. Wittkower's collection *Facebook and Philosophy*, whereby Facebook's media environment is just treated as another example through which traditional moral issues—principally posed before individual, self-sovereign human users—can be examined, even though the editor acknowledges that "what is so valuable about Facebook [is] the indeterminate meaning of so much of what it is, and what it does."[8] Indeed, this acknowledgment is followed by the conclusion that such indeterminacy allows us "plenty of space to make things mean *what we want them to*."[9]

Wittkower's playful and engaging book deserves closer examination precisely because it encapsulates some of the key issues involved in how ethics and, more broadly, philosophy travels across academic and nonacademic domains, but also in how media issues and platforms play a key role in shaping the dominant moral discourse in society. Looking at the ever more prevalent practice of social networking, the contributors to *Facebook and Philosophy* present these ethical issues as being predominantly the question of individual choice. This includes questions of how, and to whom, we present ourselves;[10] of authenticity—"authentic selves, authentic relationships, and authentic communities";[11] of truth and falsehood ("presenting a false image of who we are"[12]; of privacy, visibility, and openness;[13] of inclusiveness and participation;[14] of friendship;[15] and of bonding. For philosopher Mariam Thalos, for example, the Facebook environment "does not conduce to true bonding,"[16] because the latter is the strongest and the more authentic "face-to-face." The whole project is framed by its editor's ambition to defend, via the study of Facebook, "happiness and the pursuit of truth, beauty, creativity, and family."[17] This ties in with Maurice Hamington's attempt to think about moral problems with regard to Facebook in terms of Seyla Benhabib–inspired ethics of care—the care one as a user has for his or her Facebook friends, but also the care, or lack thereof, on the part of one's Facebook friends toward him or her.[18]

The Facebook environment thus simply seems to present us with "age-old, human problems" in all their cross-cultural and transhistorical applicability. This supposed universality of moral issues Facebook users are faced with, the kinds of problems we already supposedly knew "before Facebook," allows the contributors to Wittkower's volume to borrow rather freely and safely from the treasure chest of moral values such as the Aristotelian notion of *philia* or the feminist notion of "care" in order to evaluate Facebook's viability as a space for enacting, experiencing, and generating what philosophy has traditionally referred to as "good." The majority of the authors seem concerned with assessing to what extent the particular online practices allow for growth and flourishing, yet it is first of all human users—namely, members of Facebook's networked environment—that are supposed to grow and flourish, not the network itself. Such analyses rely on a pretechnical idea of the human, and of the interhuman (face-to-face, real, authentic) relationship to make judgments about the value of Facebook interactions. The face-to-face trope is really made prominent by the contributors to Wittkower's volume, as it stands for intimacy and privacy,[19] for palpability, preciousness, civility, and self-regulation,[20] as well as for exposure and vulnerability.[21] Though no doubt raising some interesting questions about the philosophical trials and tribulations in online environments, it is significant that the authors in their majority are not prepared to consider the possibility that those relations within the network, and thus the network itself—of which its human users are part—are perhaps themselves undergoing a radical transformation in ways that cannot be judged in advance by using ready-made moral criteria. Indeed, no matter what rearrangements of the technological setup of the world are taking place, the world is seen here as retaining a certain core identity that in turn demands some core values, with Facebook here being seen as just a new "context" to which we can apply our previously worked out moral principles. Also, the authors seem to have a strong sense of "the real," which is seen as somehow superior and more authentic to the world of Facebook. In such articulations, Facebook is reduced to a Platonic cave—sheltering us from the real world and offering only mirages of authenticity, mirror reflections of true world activities, and thus of what could "truly" count as a "good life" (or, as Condella puts it, "personal happiness").[22]

We have discussed the limitations of value-based moral frameworks for thinking about media and culture elsewhere;[23] our discussion of Wittkower's book is aimed to highlight some of the key limitations of such an approach to ethics. Yet there is also another reason why we are concentrating on this particular volume here. In its attempt to say something about an emergent media platform, *Facebook and Philosophy* curiously repeats the familiar gesture of humanizing technology by reducing it to the question of (human) use, which we discussed in chapter 1—with Facebook seen as, for example, just a tool for communication, networking, and fostering the democratic process.[24] This humanizing gesture is perfectly encapsulated by Wittkower's pronouncement: "Facebook, for the most part, is people."[25] At the same time, this forgetting of the technological side of Facebook, or what we could call *the demediation of the medium—a counter-McLuhanian maneuver that reduced media to their content—can actually take place only via other media: in this case, a print-on-paper book, using the communication medium of language.*

We Are All Facebook

Indeed, the ontology of Facebook is always inevitably mediated via other media: films (*The Social Network*, *Catfish*), books (Wittkower's volume, David Kirkpatrick's *The Facebook Effect: The Inside Story of the Company that Is Connecting the World*), newspapers and magazines (*Time*), the Internet with its various blogs, Twitter comments, Google and other search engines, as well as by Facebook itself—which constantly remediates its own form and content, as well as the form and content of other media, such as school yearbooks, diaries, blogs, and graffiti walls. This latter point is cogently argued by Ian Bogost, who in his contribution to Wittkower's volume claims that "Understanding Facebook . . . requires us to reflect partly on how it alters, suppresses, and revives earlier media."[26] He goes on to suggest that "Adopting McLuhan's stance on media ecology invites us to see Facebook as a set of media properties that both stimulate and diminish earlier media, rather than as a delivery system for content like text blurbs, photos, status updates, and applications."[27] Bogost touches upon something rather important here: *far from functioning as a mere tool for communicating, sharing things and making the "real" world more democratic, Facebook could be understood more accurately as a dynamic environment that actively shapes the spatiotemporal continuum of which we are part.* With its constant flow of data, its shaping of human and nonhuman experiences and events, and its reworking of what we understand as a "relationship" and a "connection," we could perhaps go so far as to suggest that *Facebook is a modulation of "life itself,"* if life, following Bergson and Deleuze the way we have been in this book, is duration, movement, change. Bogost seems to be drawing on this very same sensibility when he suggests, "Life events never really exist as moments in time for Facebook, but only as flows through the present."[28] His claim for understanding Facebook as a flow of duration is justified through looking at it as an inherently unstable system: one that, in the words of cyberneticists Humberto Maturana and Francisco Varela,[29] remains closed on the level of organization but not on the level of structure. We could thus say that Facebook lends itself to being understood as a cybernetic system because it preserves its dynamics, or the relationship between its parts or particles, but not necessarily all the actual particles that constitute it. These different bits, particles, and data traces constantly get incorporated into and dropped from what is understood as "Facebook." Bogost explains this process in the following terms: "Facebook amplifies the newness of what has happened recently by displaying this information first and by allowing older items to flow off the page. Nowness is encouraged on Facebook, so much so that individual moments transform into overall flow—a *feel* of now now now."[30]

If life is indeed a flow of duration and "a current sent through matter,"[31] then it is perhaps more productive to analyze Facebook as simultaneously a process and an entity. It is both a propulsion force and a system of relations that arranges animate and inanimate matter into different entities: bodies, relations, protocols, digital marks on the ever-moving "wall" or "timeline." If we are prepared to go along with such a networked understanding of the media environment in which human and nonhuman entities are not preconstituted wholes that only come together for online interaction, with a view to performing some tasks and achieving some goals, but are rather seen as mutually intra-acting, then what kind of ethical framework would we need to apply to it given that the traditional humanist value-based ethics will surely be found wanting in this context? Or, to put it differently, what kind of ethics would we need to apply to thinking about the media if our principal concern were not whether Facebook (or another "medium") was a force of good for the human but rather that we are all on Facebook—regardless of whether we have an actual account on it, or even that, in the digital media environment of today, "we are all Facebook"?

This latter proposition has been put forward by Trebor Scholz, who, in his aptly titled article "Becoming Facebook," argues that "The 'Web 2.0 Ideology' works through us, not at us. . . . We are marketing our lifestyle to each other—the books we read, the restaurants we go to, the films we watch, the people we admire, the music we listen to, the news we think is important, and even the artworks that we appreciate. It is in this sense that we are not merely 'on' Facebook but that we are becoming Facebook. . . . We are the brand."[32] Scholz highlights here the ongoing constitution of human life as not only technological and networked, but also as branded; incorporated into the apparatus of technocapitalist production in which the self functions as its own designer, distributor, marketer, and publicist. This ongoing process of human coevolution with technologies and media (which we discussed under the rubric of "originary technicity" in chapter 1) takes on unique characteristics under digital capitalism while also throwing up some new political and ethical challenges. In this light, the ethics of mediation as we see it has to address three fundamental questions:

(1) The question of biopolitics, which focuses on the production, incorporation, and management of human and nonhuman life in media networks;

(2) The question of the political economy of the media, which addresses the positioning of its human and nonhuman subjects within this economy;

(3) The question of the bioethical cut to the flow of mediation, which considers (as already suggested in chapter 3) what it means to "cut well," and ponders who will be an arbitrator of such "good" cuts.

The Biopolitics of Mediation

Let us first of all address the question of biopolitics. Admittedly, biopolitics has received so much attention in media scholarship over the recent years[33]—being simultaneously expanded to incorporate "the management of life" (and thus "everything") and contracted to stand for negativity, constraint, impasse, and ultimate lifelessness[34]—that it may seem unwise to bring this notion back in order to say something about life and the media. Yet the concept of biopolitics offers us something unique if we want to understand what is really meant by this dual process of the constitution of life as technological and its incorporation into the networks of technocapitalist production. This dual process of constitution and incorporation, of generativity and constraint, in turn calls for a closer investigation of the interlocking of the technical and biological processes of mediation, in all their different guises and forces. From this perspective, life itself participates in and unfolds through the complex and dynamic processes of mediation that are in operation at a biological, social, and political level in the world. This approach can offer us insight into what we earlier called *the vitality, or lifeness, of media*. By the vitality of media, we understand the possibility of the emergence of forms ever new, or the potentiality to generate unprecedented connections and unexpected events. *It is precisely in the tension between the generative potentiality of mediation and the constraining aspects of the particular media setups and networks that biopolitics as a historically specific management and regulation of life presents itself in full force.* With this proposition about *the lifeness of media, or their underlying mediation*, we are bringing together what Eugene Thacker has identified as two conceptualizations of life in Western philosophy: a "biopolitical" one, whereby life is seen primarily as form, and a "phenomenologico-affective" one, with life understood as time.[35] Such a materialist positioning of life as both form and process aims to foreground life's immanent, productive dynamism.[36] But this productive dynamism is subject to singularities, moldings, and modulations—by economic, political, and technical forces.

The Neurological Turn?

According to some theorists of technology, the Internet is one area where this interlocking of the technological and the biological on the level of life is occurring most forcefully. By "the Internet," we understand here not just the global system of computer networks but rather the whole technohuman ensemble consisting of servers, individual computers, cables, data flows, individual platforms, websites, and social networking portals such as Facebook, Twitter, and Amazon as well as their human users. N. Katherine Hayles argues that "networked and programmable media are part of a rapidly developing mediascape transforming how citizens of developed countries do business, conduct their social lives, communicate with one another, and—perhaps most significant—think."[37] Drawing on current research in neuroscience, she goes so far as to suggest that we find ourselves in the middle of "a generational shift in cognitive styles"[38] that entails "the neurological re-wiring" of the human brain as a result of performing small repetitive tasks—a process that neuroscience ascribes to the brain's "plasticity" (also called "neuroplasticity").[39] "In contemporary developed societies, this plasticity implies that the brain's synaptic connections are coevolving with an environment in which media consumption is a dominant factor," claims Hayles.[40] It also is in this sense that being on Facebook should therefore rather be understood as "becoming Facebook." Indeed, *Facebook participation is not just a social experience but also a biological one: it is something that engages our cognitive apparatus, possibly adjusts our "plastic brain," connecting as it does our eyes, fingers, and minds to the expansive network architecture of the web.*

This conclusion in itself is not that novel—we can hear in it echoes of Marshall McLuhan's declaration that "the medium is the message," as well as his emphasis on the physicality of media technologies—but it has acquired a new inflection under the aegis of what some are calling "the neurological turn."[41] Nicholas Carr, author of publications beloved by mainstream media, with titles such as "Is Google Making Us Stupid?" and *The Shallows: What the Internet Is Doing to Our Brains*, argues:

> Even as the Internet grants us easy access to vast amounts of information, it is turning us into shallower thinkers, literally changing the structure of our brain. . . . The Net's ability to monitor events and send out messages and notifications automatically is, of course, one of its great strengths as a communication technology. We rely on that capability to personalize the workings of the system, to program the vast database to respond to our particular needs, interests, and desires. We want to be interrupted, because each interruption—email, tweet, instant message, RSS headline—brings us a valuable piece of information. To turn off these alerts is to risk feeling out of touch or even socially isolated. The stream of new information also plays to our natural tendency to overemphasize the immediate. We crave the new even when we know it's trivial.[42]

Determining the scientific accuracy of such applications of neuroplasticity to the understanding of media coevolution is not our primary concern here, and many have already raised serious questions about Carr's diagnosis.[43] Indeed, as philosopher Catherine Malabou argues in her book *What Should We Do With Our Brain?*, in spite of all the developments in neuroscience over the last fifty years, including the observation of the brain in vivo thanks to MRI and PET technologies, research on the reconstitution of the bodily schema via an inclusion of foreign members after an organ transplant, and increased understanding of neurodegenerative disorders such as Alzheimer's or Parkinson's, "neuronal man still has no consciousness."[44] We do not yet fully understand what this neural plasticity actually *means* for us and for the world, and—worse—"given that the activity of the central nervous system, as it is revealed today in the light of scientific discovery, presents reflection with what is doubtless a completely new conception of transformation, we nonetheless have the feeling that nothing is transformed."[45] Another way of putting it is that even though our brain is "plastic," we cannot grasp the true consequences of such plasticity or envisage any radical sociopolitical change, more often than not reducing any potential transformation, of both the brain and the world, to the already known (or what Bergson termed "solids").

Malabou is therefore rather skeptical of what we could describe as "brain-hype"[46] and, in particular, of the reduction of the brain's plasticity to "flexibility" understood in market and labor terms—which "purely and simply coincide[s] with the spirit of capitalism,"[47] and which actually lacks "the power to create, to invent," in spite of its all-pervasive rhetoric of innovation.[48] She insists that "the fundamental, organizing attribute of plasticity" is "its power to configure the world,"[49] yet at the same time she points out that "plasticity is also the capacity to annihilate the very form it is able to receive or create,"[50] not just produce new synapses and new opportunities ad infinitum. "The neurological turn" is therefore arguably an aspect of what Malabou describes as a "reductionist discourse that models and naturalizes the neuronal process in order to legitimate a certain social and political functioning."[51] Let us make it clear that Malabou is not against science, only against scientific reductionism. Indeed, her argument is that the actual plasticity and hence the vitality of the brain gets bracketed and foreclosed in such pronouncements. She derives this argument from the thought of Bergson, who has shown "that every vital motion is plastic, which is to say that it proceeds from a simultaneous explosion and creation. Only in making explosives does life give shape to its own freedom, that is, turn away from pure genetic determinism."[52]

We want to follow in Malabou's footsteps and pick up on *the "neurological turn" trope as a particular discursive articulation of the current media experience with reference to scientific (or sometimes pseudoscientific) concepts*. Under the guise of "real science" such an articulation arguably depoliticizes the analysis of the technological domain by passing off cultural trends and tropes as scientific facts. As Geert Lovink explains in his pointedly titled piece "MyBrain.net":

[Nicholas] Carr and others cleverly exploit the Anglo-American obsession with anything related to the mind, brain and consciousness—mainstream science reporting cannot get enough of it. A thorough economic (let alone Marxist) analysis of Google and the free and open complex is seriously uncool. It seems that the cultural critics will have to sing along with the Daniel Dennetts of this world (loosely gathered on edge.org) in order to communicate their concerns.[53]

True to his word, Lovink then proceeds to outline a politicoeconomic critique of the literalization of the old "time is money" adage, whereby online "duration" and "instantaneity" are being turned into commodities via social networking platforms. He goes on to argue:

The pacemaker of the real-time Internet is "microblogging," but we can also think of the social networking sites and their urge to pull as many real-time data out of its users as possible: "What are you doing?" Give us your self-shot. "What's on your mind?" Expose your impulses. Frantically updated blogs are part of this inclination, as are frequently updated news sites. The driving technology behind this is the constant evolution of RSS feeds, which makes it possible to get instant updates of what's happening elsewhere on the web. The proliferation of mobile phones plays a significant background role in "mobilizing" your computer, social network, video and photo camera, audio devices, and eventually also your TV. The miniaturization of hardware combined with wireless connectivity makes it possible for technology to become an invisible part of everyday life. Web 2.0 applications respond to this trend and attempt to extract value out of every situation we find ourselves in. The Machine constantly wants to know what we think, what choices we make, where we go, who we talk to.[54]

This "nowness" we discussed earlier on in this chapter (as well as in chapter 4), which is encouraged on Facebook, Twitter, and other social networking and microblogging platforms (so much so that, as Bogost puts it, "individual moments transform into overall flow—a *feel* of now now now"),[55] becomes here a flow of capital. In the process, human lives are being immediately even if not directly incorporated into the production and distribution of technovalue. As carriers and signalers of valuable information, such platforms turn not just us as their users but also "life itself" into a product, constantly remolded and repackaged via the flickering pulsations and pings of data, with Twitter's original, "What are you doing" replaced in November 2009 by the more impersonal, less human-centered "What's happening?"

Yet by no means do we want to suggest that biological influences do not matter, or that any attempts to incorporate the neural into academic analyses of the media should be countered with a one-way return to more analytical and interpretative humanist frameworks, especially that, as psychologist Steve Pinker claims—in a way not dissimilar from Malabou's philosophical argument—the talk that "experience changes our brain" is both fundamentally obvious and overdetermined. He admits that "cognitive neuroscientists roll their eyes" at such formulations, as "every time we learn a fact or skill the wiring of the brain changes; it's not as if the information is stored in the pancreas."[56] However, Pinker also points out that "the existence of neural plasticity does not mean the brain is a blob of clay pounded into shape by experience."[57] *What is important for us is therefore not so much the recognition that the media may be affecting "us," but first of all the acknowledgment of the mutual co-constitution of "media" and "us" along both cultural and neural lines, that is, the intertwined process of media coproduction.* Another way of putting is that, to cite Bernard Stiegler, "the psychic apparatus is continuously reconfigured by technical and technological apparatuses and social structures."[58] The recognition of the mutually constitutive aspects of "the brain" and "the world," of media and mediation, will allow us a way out of the familiar nature-culture dualism while also providing an impetus to develop some new theoretical frameworks which seriously engage with such "naturecultural" entanglements.

The Political Economy of the Media, Otherwise

The recognition of this media(ted) co-constitution will inevitably affect the ethical frameworks and questions through which we can understand both media and ourselves. If life is mediation and if mediation inscribes itself in the biopolitical framework in which life is being modulated on the technical and corporeal level, then arguably the ethics of mediation must be a form of bioethics, an ethics of lifeness rooted in the entangled notions of differentiation and becoming, with life studied in both its molecular and social aspects. [59] However, in order to avoid subscribing to an ahistorical metaphysics of life in which life is primarily understood in spiritual terms,[60] any attempt to consider the bio/ethics of mediation has to engage seriously with the political economy of the media that shape the landscape of mediation. The political and economic issues raised by this mediation of life—that is, the emergence of life with and via the media—throws up a number of ethical problems, the fundamental one of which derives from the very business model of many social networking sites such as Facebook, a model "which is based on the exploitation of networks of alleged friendship to sell stuff."[61] The marketing side of Facebook is made relatively transparent (see chapter 4), with the site's users accepting "the implicit bargain that they get to connect via a really 'cool' platform in return for making themselves available to marketers. They don't buy Facebook, it sells them," as Foster puts it.[62]

Stiegler's *For a New Critique of Political Economy*—a book that provides a rigorous analysis of the entanglement of the neural and economic, and the libidinal and the cultural, and that is partly indebted to Hayles's work on shallow attention discussed previously (and taken up in much more detail in Stiegler's *Taking Care of Youth and Its Generations*)—can offer a helpful response to the situation described earlier. It contains a series of theses aimed at dealing with what Stiegler terms "the first planetary economic crisis of the capitalist industrial world,"[63] and which we discussed under the rubric of the "credit crunch" in chapter 2. Like us, Stiegler is referring here to the global economic situation in the early twenty-first century, when the consumerist model has reached its limit, and when investment has turned into speculation. The emergence of "cognitive" capitalism, whereby labor, as shown in chapter 4, has become to a large extent "affective," has led to the exteriorization of memory, in all its forms, by technology. (To give just two quick examples, people on the whole tend to remember far fewer phone numbers than they used to, with many not even being able to recall their own number, and the loss or accidental deletion of an electronic diary—which is often "shared" with managers, work colleagues, and others—leads to panic resulting from the inability to recall one's order of the day, scheduled appointments, and deadlines.) Memory today is thus becoming "the object of sociopolitical and biopolitical controls through the economic investments of social organizations, which thereby *rearrange psychic organizations* through the intermediary of mnemotechnical organs, among which must be counted machine-tools . . . and all automata—including household appliances [such as talking toasters or malicious doors from our chapter 4], as well as 'the internet of things.'"[64] Consequently, any present critique of political economy must for Stiegler engage "in a critique of libidinal economy";[65] or, to put it differently, if we want to understand the working of capital, we must also study the working of desire, because financial and corporeal intensities reinforce each other. In response to this diagnosis, Stiegler offers a different model of investment—one premised on common social political goals, or what he terms a "common desire,"[66] rather than on private capital and individual satisfaction of wants and needs.

A Poison and a Cure

For Stiegler, the present economy is a "pharmacology,"[67] which is to say that it displays the mixed properties of Plato's *pharmakon*: it is both a poison and a cure. In *Phaedrus*, the Greek philosopher originally describes writing as a tool that has curative properties (*pharmakon*), only to then reposition it as a substance that poisons truth, memory, and the soul (again, *pharmakon*). In other words, the current economy, with its libidinal flows of technology, capital, and desire, potentially strengthens life and facilitates the production of the new (new ideas, investments, conjunctures, products). Yet it also produces blockages on both a corporeal and a global level, ultimately harming the bodies that are supposed to nourish it and ensure its "good health." The working of the pharmacological economy of this kind does not oppose remedy and cure in a dialectical manner; it is rather a "*composition* of tendencies."[68] Hence, it would be too simplistic to see, say, Facebook as *just* exploitative, immoral, and posing a threat to our privacy and sovereignty without recognizing its generative potential. Indeed, for Stiegler this composition of tendencies within the current economic dynamics also facilitates the development of a "system of care," which potentially allows for the individuation of the human on three different levels: technical, psychic, and collective. This tension between poison and cure is important here. Indeed, such potentially *productive* economy of individuation is inextricably interwoven with what Stiegler terms "a *dis-economy* of *pharmaka*," which is the name he gives to "what results from the appearance of any new *pharmakon*" when it disables psychic and collective individuation.[69] (Or, to put it in less philosophical terms, *it's good, until it's bad*.) Stiegler then concludes, somewhat surprisingly and overdecisively, that this is precisely what is "occurring today with the technologies of 'social networking,' for which no political economy and no system of care is prescribed by any public authority; or again, it is what occurs in the course of the synaptogenesis of the infantile cerebral organ when the audiovisual short-circuits the transitional object, the infantile psychic apparatus being thereby proletarianized."[70] Stiegler thus seems to be subscribing all too quickly to the previously discussed logic of the neurological turn, whereby a clear causative link is posited between a phenomenon in the world (social networking portals, audiovisual media) and infants' brains. This kind of repudiation of certain cultural forms and practices as inherently damaging is of course familiar to scholars in media, communications, and cultural studies. It has been explored most productively under the rubric of the "moral panic"—a term used to describe the "mounting of a symbolic crusade"[71] against a perceived threat in a society at a given moment in time. Stiegler's reaction can thus perhaps be seen as an unwitting attempt to rouse a moral panic for the "neurological age," whereby the current "folk devil" (television, the Internet, Facebook) is said to be posing a threat not so much to the social order but to something deeper and more intrinsic: the human brain. The turn to neuroscientific vocabulary works to disguise an attempt to create a moral panic as a factual description of the event.

Yet if the politics of today can be understood most productively in terms of biopolitics—a point we have argued previously in this book and one that Stiegler also puts forward in *For a New Critique of Political Economy*—then we can conclude that the pharmaceutical discourse of his economic theses is not out of place, even if we remain suspicious of his a priori denigration of particular media forms and platforms. Indeed, his politicophilosophical proposition is aimed at bringing about a remedy to a situation in which the potentially harmful technical-corporeal setup can harness the very energy inherent in the system and put it to more generative, productive uses, by "intensifying individuation"[72]—but also, perhaps (to follow Malabou), by mobilizing the brain's vitality to reconfigure the world. Such management of the flow of life is nothing new: through history we have been developing ways of managing it via processes of what Stiegler calls, with a nod to Husserl, "tertiary retention," with the gestures of the body gradually becoming "analytically reproducible" in the form of digits, letters, and other symbolic marks—a process that has resulted "in retentional grains that one can call *grammes*. . . . And this is why we posit that the evolution of tertiary retention, from the Neolithic age until our own, constitutes a process of grammatization,"[73] claims Stiegler. Grammatization becomes here a way of providing control over the flow of life, of managing its excess, or—to use our vocabulary—of carving mediation into media. Writing, understood as the hypomnematic (that is, serving-as-a-reminder) leaving of traces, is the most prominent example of grammatization in Stiegler's work. Online networking sites can therefore be seen as spaces wherein this process of the structuration of one's psyche takes place because they allow for the emergence and reproduction of patterns, discrete units of code and meaning, time and space. Yet technologies and media will not be anything *by themselves*. This is why Stiegler postulates that a political economy and a system of care are needed if we are to avoid the proletarianization of our identity and memory as a consequence of many hours of unpaid "affective labor" spent on feeding corporations such as Facebook and Google our fantasies, desires, and customer preferences in the form of free data. Indeed, such a system of care is needed to stop us from turning into workers on a digital assembly line, constantly clicking to "friend," "poke," and "like" and thus producing ever more network growth.

Cut Again. Cut Better.[74]

Drawing on Deleuze, Malabou offers an interesting reading of this process of grammatization, or making gaps and cuts in the flow of life—a logic that for us also underpins the relation between media and mediation—when she writes:

Cerebral space is constituted by cuts, by voids, by gaps, and this prevents our taking it to be an integrative totality. In effect, neuronal tissue is discontinuous: "nerve circuits consist of neurons *juxtaposed* at the synapses. There is a 'break' between one neuron and the other." Between two neurons, there is thus a caesura, and the synapse itself is "gapped." (One speaks, moreover, of "synaptic gaps.") Because of this, the interval or the cut plays a decisive role in cerebral organization. Nervous information must cross voids, and something aleatory introduces itself between the emission and the reception of a message, constituting the field of action of plasticity.[75]

If grammatization is in itself a process of managing the flow of life and time by making insertions and cuts into it, the bio/ethics of mediation can therefore be understood as a way of making "good" cuts. Such bio/ethics, although not a panacea, can perhaps serve as a form of therapeutics desired by Stiegler, a nonnormative framework of care that will be able to take us out of economic and libidinal stasis. This kind of bio/ethics will safeguard human individuation with the media (which entails as much singular instants of becoming-individual as it does the indivisibility of the human from his or her mediation), but it will also protect *the vitality, or lifeness, of media.* We are therefore prepared to award the technologies of "social networking" more generative potential than Stiegler is—for whom they unambiguously find themselves on the side of negentropy because they are not assisted with enough systemic regulation and care. Yet the very bodies, minds, and machines incorporated in the social networks are already potentially a source—although not a guarantee—of the exertion of good care, or an enactment of "good cuts." Bio/ethics becomes here such a way of enacting good cuts, that is, cuts that facilitate the flow of life through the network without drowning us in the process.

As argued in chapter 3, cutting is fundamental to our emergence in the world, as well as our differentiation from it. It thus has an ontological significance: it is a way of shaping the universe and shaping ourselves in it. Through cutting, we become individuated, or temporarily stabilized, as "selves" in the process of engaging with matter and attempting to give it (and ourselves) form. By making incisions into material reality—with our eyes, our bodily and cognitive apparatus, our language, our memory—we instantiate separation and relationality as the two dominant vectors of our locatedness in time. Bio/ethics therefore becomes a practice of time management, a speed control mechanism aimed at regulating the "geographical, biological, demographic and psychic systems [that] find themselves disadjusted, leading to their disequilibrium, rather than to beneficial disruptions."[76] It is also a framework for stopping the intellect—which, according to Bergson, is unable to "think true continuity, real mobility"—from cutting things too quickly, and thus foreclosing on the creative duration of life. On the other hand, we should pay heed to Malabou's warning against reducing the cut that separates the neuronal from the mental, or naturalness and intentionality, to synaptic gaps, and therefore supplementing or "relaying" scientific explanation with "*interpretation.*"[77] "Plasticity, rethought philosophically, could be the name of this *entre-deux,*" she adds. Indeed, it could serve as a name for a very particular form of relationality, one that combines the entanglement and mutual becoming of the self and the world with an injunction to make this becoming happen, to make cuts to the duration of things.

Rather than lead us to a moralistic rejection of social networking as inauthentic, antisocial, and artificial, a bio/ethical approach to mediation can help us see Facebook, Twitter, and other such platforms and activities as constitutive flows of mediation that are instrumental in our becoming in the media world (even if our attitude to, or position on them is one of rejection, denial or neutral noninvolvement). A bio/ethics that remains attuned to the temporal fluidity of media calls on us to develop interventions from the midst of the field of life forces, so to speak. It can also equip us with a series of techniques and strategies for enacting what Karen Barad has termed "agential cuts," which "enact a resolution *within* the phenomenon of the inherent ontological (and semantic) indeterminacy."[78] All of this casts a different light on Andrew Keen's somewhat hysterical assessment of the supposed totalitarianism of the digital world—a setup also described in terms of the "connective tissue of society" by Clay Shirky or as the new "nervous system of the planet" by Hilary Clinton, whereby "in an age of transparent online communities such as Twitter, LinkedIn, and Facebook, the social has become, in Shirky's words, the 'default setting' on the Internet, transforming digital technology from a tool of our 'second lives' into a central part of real life."[79] This apparent repudiation can perhaps be seen instead as a recalibration of media anthropocentrism that typically positions the human as both the producer of media and their key recipient.

Significantly, articulations of media in terms of processes of mediation have been recently proliferated outside, and perhaps also ahead of, the academic context. In a clear repudiation of the idea that digital data and, by extension, digital media are discrete entities that can be safely kept in one location, Bradley Manning, a US Army private who broke into the military databases and passed on the secret information to WikiLeaks in 2011 and who is said to have had links with the creative Boston hacker culture, brought up in conversations with his computer researcher friend from MIT, David House, the notion of "'neuro-sociology,' the idea that the human race is now connected by the Internet, which is like a nervous system for the human race enabling people to organize much quicker and faster." This concept led House to pose the following question: "What does that do to us as a species from an anthropological point of view?"[80] In a similar vein, on attending the 2011 South by Southwest festival of film, music and technology in Texas, British journalist Oliver Burkeman announced that "the Internet was over": not in the sense that it was not important any more but rather that it extended well beyond discrete and identifiable computers, servers, cables, and users to become "kind of everything," with boundaries between the physical and the virtual being no longer sustainable. We can see this hyperbole as a philosophical rearticulation of the earlier notion of ubiquitous computing and of the early metaphors of the Internet itself as a global brain, central nervous system, or evolving superorganism whereby a clear parallelism is established "between the transformation of the spirit of capitalism (between the sixties and the nineties) and the modification, brought about in approximately the same period, of our view of cerebral structures."[81] And yet such an articulation, even if indeed wedded to the rather problematic economic discourse that promises flexibility, connectivity, and growth, can at the same time be seen as an attempt to think beyond the spatialized framework that understands media principally in terms of their locatedness and boundedness. Burkeman himself is aware of the grandiosity and overdetermination of such claims and of their hyperbolic nature and is not so much interested in providing a totalizing narrative about the current state of technology or in reducing it to a sound bite. Instead, he traces the possible openings within the media landscape by looking at the working of what could be termed the apparent forgetting of technology in a number of different projects. He writes: "When the GPS system in your phone or iPad can relay your location to any site or device you like, when Facebook uses facial recognition on photographs posted there, when your financial transactions are tracked, and when the location of your car can influence a constantly changing, sensor-driven congestion-charging scheme, all in real time, something has qualitatively changed. You're still creating the web, but without the conscious need to do so."[82] This ties in with Tim O'Reilly's observation: "Our phones and cameras are being turned into eyes and ears for applications."[83]

The extension of media toward the complex processes of mediation into which the human is interwoven and from which he or she also emerges *as human* shifts the level of analysis from looking at singular subjects and objects (film directors, journalists, bloggers, viewers, cameras, TV sets, computers, cables) to that of multiscalar processes of media emergence. Even though the agential resolution, the production of subjects and objects—including humans and media—is not something that remains fully under human control (indeed, that would be a logical impossibility), the ethical dimension of what we earlier called "making good cuts" certainly does, at least to some extent. It is this domain of human control from within the midst of life forces, restricted and overreaching as it is, that provides a domain for ethical work. *Bio/ethics thus names those processes of agential resolution that carry a human inflection: they are processes of "differential cutting," of making pragmatic in-cisions into the flow of life that also have the force of ethical de-cisions.*

Revisiting Stiegler's reading of the myth of Prometheus and Epimetheus as a story of the human emergence with technology (which we discussed in chapter 1), Federica Frabetti comments on this tragic fate of the human—who, as a result of Epimetheus's "forgetting," has been deprived of "qualities" such as wisdom and the ability to live productively with one another. Put in charge of his own fate, the human lacks finality: the ability to make prudent decisions about his life and future.[84] Frabetti goes on to argue the following:

> Stiegler's reworking of the myth clearly demonstrates how for him technology raises the problem of decision, and how this encounter of the human with decision in turn constitutes time—or rather, what Stiegler calls "technical time." Technical time emerges because human beings experience their capacity for making a difference in time through decisions. Temporality is precisely this opening of the possibility of a decision, which is also the possibility of creating the unpredictable, the new.[85]

Even if decisions are characteristic of all actual entities, not just humans—as argued by Whitehead and Steven Shaviro[86] after him—in the sense that all beings are capable of selecting from among different behavioral options and performing truly novel acts that do not require consciousness, the (at least partially or reflexively) conscious temporalizing, grammatizing decisions made by the human arguably deserve attention. This is not to return to a humanist standpoint and uncritically proclaim the superiority of the human—but to point to the singularity of this perhaps narrow and rare section of the decision spectrum in the universe where their enactment requires a certain mobilization of a cognitive and neurocortical apparatus (as well as, to cite Derrida, a leap of faith). Interestingly, Shaviro[87] suggests that rather than perceive consciousness and responsibility[88] as the grounds or preconditions for decision or the exercise of free will, we should see them as the *consequences* of making decisions and (thereby) exercising free will. The emergence of both out of the plethora of actions undertaken by living beings, actions that go under the name of "decisions," would thus open the door to this narrower category of responsible and (at least partially) conscious decisions made by the temporarily stabilized human, carved out of the wider network of life forces. It is this second category of decisions that is the focus and subject of our bio/ethics understood as an ethics of mediation.

Responsible decisions about life, made from amid life itself, will focus on the conservation of life, or what Stiegler calls "retention,"[89] in a way that does not quash life's generative potential. Life is posited here as a minimal value—a basic condition for the existence of the human—but not necessarily as an ebullient and overbearing force that will always "find a way." *An ethics of mediation can thus be perceived as a therapeutics because it needs to take care of life and its various incarnations, while also allowing for the decoupling of life's generativity from neoliberal productivity.* The key ethical question facing us in the era of Facebook, Twitter, and other social networks is therefore not whether we should or should not be "on them," but rather how to emerge with social media, and how to become-different from them.

7 Remediating Creativity: Performance, Invention, Critique

Reclaiming Creativity

The principal aim of this chapter is to reclaim creativity from many of its current contexts—predominantly those associated with Romanticism, neoliberalism, and what we may call, for short, "Deleuzianism" (although, let us make it clear right from the start, Deleuze is not our enemy here). Drawing on Henri Bergson's idea of creativity as developed in his book *Creative Evolution*, we shall continue working with the notion of "mediation" as a device for thinking about media events and inventions. Mediation stands for a set of entangled economic, cultural, social, technical, textual, and psychological processes through which a variety of media forms develop in ways that are at times progressive and at times conservative.[1] In other words, as we argued in chapter 2, mediation is always generative, but it is not necessarily progressive. A discussion of creativity will therefore require a theory of decision if we are be able to make judgments about the progressivism and conservatism of various media inventions. Even though it remains in dialog with creative arts, including those that make use of and produce various forms of (new) media, our attempt to reclaim creativity is thus not just an aesthetic endeavor: it also carries with it an ethical injunction to "invent well."

Any attempt to bring back "creativity" to the study of media needs justification, given the various uses this term has been put to by artists, policy makers, and scholars worldwide—including those in media studies. Creativity is a rather nebulous concept encapsulating all these human characteristics and skills—seen by some as innate and by others as learned—that involve the ability to see things in a new way; to invent new ideas, concepts, and objects; and to provide new associations between existing entities. One can hear in it an echo of the Latin word *creatio*, with its associations of begetting, or "creation from nothing"—in the way God is supposed to have created the world. The term is often linked with originality and with breaking new ground and is variously explained by genius, divine intervention, or chance. Due to its conceptual openness, creativity has been claimed by various disciplines and fields of enquiry: arts and humanities, architecture and design, education, science, economics, and business.

The incorporation of the language of creativity by the international business-politics nexus under the umbrella of so-called creative industries is of particular concern to us here, partly because of the role "new media"—as both a set of discourses and material practices—have played in this process of incorporation. This process has involved a whole series of disciplinary mediations, including "*the conceptual and practical convergence* of the CREATIVE ARTS (individual talent) with Cultural Industries (mass scale), in the context of NEW MEDIA TECHNOLOGIES (ICTs) within a NEW KNOWLEDGE ECONOMY, for the use of newly INTERACTIVE CITIZEN-CONSUMERS."[2] In the words of John Hartley, an Australian academic and one of the most vociferous supporters of the "creative industries" paradigm, at a particular point in the history of Western postindustrial capitalism creativity was repositioned as "the driver of social and economic change."[3] Playing a key role in the emergence of a wholly new mode of production, creativity became a marketable and transferable commodity for the information age in the same way that cars had been in the Fordist era and coal in the early industrial period. Although the idea of the creative industries was introduced in Australia by Paul Keating's government at the beginning of the 1990s, it really came into its own in Tony Blair's 1997 "Cool Britannia," whose "zealous modernizers renamed the Department of National Heritage as the Department of Culture, Media and Sport (DCMS), and promoted, as its bailiwick, a paradigm of self-directed innovation in the arts and knowledge sectors of the economy."[4] The model of creativity as a product in itself allowed for it to be exported to countries as diverse as Russia, Brazil, Canada, and China. The situation in the United States was somewhat more complicated, as explained by Andrew Ross, due to the (not always maintained yet explicitly claimed) separation between government and cultural policy. Less preoccupied with incorporating creativity as part of their agenda and more with securing ownership rights over every domain of cultural expression, "in the US, the creative industries are more routinely, and bluntly, referred to as copyright or IP industries."[5]

The economic and political limitations of the "cultural industries" paradigm have been well argued and carefully documented, not only by Ross but also by other cultural critics such as Angela McRobbie, David Hesmondhalgh, Geert Lovink, and Ned Rossiter,[6] with such a repositioning of "creativity" seen as leading to the proletarianization of the newly emergent "creative class," the demise of viable independent cultural production, and the increased individualization of the creative labor force. This creative labor force is increasingly involved in what is being termed an "affective" form of labor.[7] In the context of the previous arguments, it may seem risky or even imprudent to attempt to reclaim creativity as a viable strategy for thinking about the media differently. Yet it is precisely against the all-consuming power of creative industries' narratives and practices that we set out with our attempt to position "creative media" as an alternative paradigm for envisaging some new ways of thinking about, and with, the media. Needless to say, we have no interest in reinstating the earlier humanist version of creativity as a human ability to create "from nothing" one driven by genius, serendipity, or divine intervention. Instead, our attempt to think about creativity turns to Bergson's "creative evolution," a theoretical proposition that, when coupled with Derrida's politicoethical philosophy, becomes for us an injunction "to invent well." Before we go on to explain how the force of this creative evolution operates and what we really mean by "inventing well" in this context, we want to signal that various efforts on the path toward reclaiming creativity from creative industries advocates have been made in recent years, particularly by theorists and practitioners associated with new media.

Creative Industries versus New Media

In the introduction to their book *MyCreativity Reader*, Geert Lovink and Ned Rossiter outline a proposal for a new "creative subject" who is "neither a citizen nor a consumer. Web 2.0 makes loud noises about the false synthesis of the so-called 'prosumer,' but this does not get us very far other than reiterating the logic of individualisation."[8] The authors then postulate "self-valorisation as a productive concept that grants legitimacy and possible stability to collaborative practice."[9] This idea is part of a bigger plan that involves an attempt to redefine creative industries beyond activities associated with the creation of intellectual property and economic value. Instead, they promote the "freedom to commit senseless acts of creativity."[10] In trying to redeem creativity from the logic of capitalist production, Lovink and Rossiter nevertheless end up reinscribing it in the former, Romantic framework of the insubordinate genius. Even if their creative subject is now collaborative rather than individualistic, it is driven by the very same desires, motivations, and fantasies that had shaped the Romantic creative: those of artistic freedom, self-worth, and individual fulfillment. In this way, they merely replace capitalist liberalism with Romantic one, albeit now dressed in the language of fluid libidinal attachments and senseless acts of productivity. The individual liberal subject is multiplied here to form a collective but not challenged in any significant way.

Such a criticism of the philosophical assumptions behind Lovink and Rossiter's project is perhaps somewhat unfair or even misguided, in that it overlooks their important political ambition: to analyze the structural injustice of the current creative industries and to outline an economic alternative for the plethora of artists, media producers and other members of "the communicative class," whose skills have been made all too transient and replaceable in the era of cognitive capitalism.[11] Similar aspirations to protect the labor practices and livelihood of many "creatives" lie behind many other sociological analyses of creative industries. Applying the methods and concerns of traditional labor sociology to the current sociopolitical conjuncture, writers such as Ross, McRobbie, and Gregg[12] have conducted ethnographic studies of creative labor, interviewing artists, designers, and "city creatives" in an attempt to make their story heard and the precariousness of their working conditions revealed. Yet although we certainly recognize the political and moral importance of research of this kind, we also hope to do something slightly different from those kinds of sociological analyses, in which economics still functions as a defining framework for creativity, with capitalist postindustrial production continuing to be positioned as a terrain for its expression, confinement, and reinscription.

Of course, this is not to suggest naïvely a move "beyond" capitalism or economics, as questions of production, calculation, and exchange are always at work within any sociocultural practice. Indeed, creativity is inevitably tied in with capitalism because, in both its monetized and nonmonetized forms—such as as music making, poetry writing, engineering, art, cooking, or knitting—it participates in the intertwined process of production, consumption, and distribution through formal and informal communicative channels and networks made up of record labels, publishing houses, manufacturing plants, computer systems, telephone exchanges, and chats with friends. It is just that in our attempt to reclaim creativity we want to shift this very term from the context of the cultural industries debates—a context that, at least in media, communications, and cultural studies, has overdetermined the understanding of creativity as a cultural concept. The discussion of creativity through the creative industries lens therefore always seems to concede—as if in advance—the ownership of this concept to the creative industries advocates. This concession seems to spring from the assumption that the only way to reuse "creativity" as an enabling term is to wrestle it away from the dominant economic framework and to outline a different—more socially aware, more collaborative, more rewarding—economy. In the majority of these projects, however, the emphasis is still very much on the *analysis* and *critique* of particular creative industries. Any suggestions of a possible emergence of challenges to the dominant paradigm thus end up reaffirming the dichotomous model of cultural industries, in which "dominant" cultural production (primarily corporations-driven, profit-oriented, and exploitative) is positioned against its "alternatives" (seen, in turn, as more personally rewarding, liberating, and sometimes also subversive toward the dominant mode).

Beyond Analysis: For a "Socially Engaged, Critical Creativity"

Again, the politicoethical significance of scholarship that analyzes and critiques cultural and media industries and their attempts to reinvent the so-called creative labor practices under the guise of collaboration, commonality, sharing, and gifting is not being contested here by us. Yet what if we were to mobilize the critical dimension of the analyses of creative media processes and products and combine it with an actual attempt to *produce creative media*, while also subjecting the notion of critique to a critique? In other words, what if, rather than just write about the production of creative media by others, we could mobilize the very media that are being critiqued as objects of creative industries' analyses and put them to critical uses, to think *with* and *through* them about change, invention, and sociocultural transformation? What if the roles of the cultural critic and the cultural producer were to be combined in this process of developing what Angela McRobbie has termed a "socially engaged, critical creativity"?[13]

"Artivist" projects that have gained the moniker "tactical media" have taken some steps toward enacting such critical creativity. For David Garcia and Geert Lovink,

> Tactical Media are what happens when the cheap "do it yourself" media, made possible by the revolution in consumer electronics and expanded forms of distribution (from public access cable to the Internet) are exploited by groups and individuals who feel aggrieved by or excluded from the wider culture. Tactical media do not just report events, as they are never impartial they always participate and it is this that more than anything separates them from mainstream media.[14]

Examples of such projects include Creative Art Ensemble's *Flesh Machine* (1997–1998), a series of performances in which participants were invited to take donor screening tests and gather information about reproductive technology as well as consider the sociopolitical aspects of such procedures; or Josh On and Futurefarmers' *They Rule* (2001/2004), a website that allows users to visualize intricate connections between Fortune 100 corporations and directors. Positioned as "media of crisis, criticism and opposition,"[15] tactical media operate via creating "disturbance" and "disruption of a dominant semiotic regime."[16] In basing the rationale for their projects on the dialectical relationship between the (bad/uncritical) mainstream and the (good/critical) oppositionality, many tactical media "hacktivists" seem to lose sight of the conditions of material and symbolic production that are involved in setting up such oppositions. Even if Garcia and Lovink suggest that tactical media should not be seen as simply oppositional because they are not set against any coherent enemy, they still position the tactical media producer as an outsider to the dominant system, thus slipping, in the words of Caroline Bassett, into the romance of subcultural resistance without being able to register the limits of this discourse.[17] For all the *critique* of creative industries, then, the hegemonic idea of creativity as defined by the latter is actually reconfirmed in many tactical media projects, even if it comes with a dialectically inverted vector when presented in its oppositional or tactical guise.

The Exuberance of Life and Its Theorists

Is it possible to invent (new) media otherwise, without falling back onto their predetermined patterns, models, and hierarchies? Deleuze encourages us to go beyond the established "schools" that regulate the creative process in the arts and media and to recapture the "creative functions" of the media themselves that transcend their "author-function."[18] For Deleuze, the key issue today "consists in reinventing—not simply for writing, but also for the cinema, the radio, the TV, and even for journalism—the creative or productive functions freed of this always reappearing author-function."[19] He sees these functions as moving beyond the constraints of the individual I. Proceeding "by intersections, crossings of line,"[20] such creative functions assemble multiple enunciations, actions, and affects, some of which may not even be human. What emerges as a result of this is "a living line" that is always inevitably temporary and broken but that can help us envisage "something else"—the truly new media as we do not perhaps know them yet. The creative impulse behind any such creative media comes to Deleuze—as much as it does to us in this book—from the writings of Henri Bergson, for whom creativity is about how reality produces effects "in which it expands and transcends its own being."[21] As a vitalist, Bergson derives this notion of creativity from the philosophicoempirical observation of living organisms of various scales and from their innate propensity for movement.

With this in mind, could we envisage extending the creative agency of the critic-producer beyond individual or even collective creators to incorporate the creative potentiality of media themselves, that is, their ability to evolve and generate forms ever new? How could we achieve this "critical creativity" without transferring all responsibility for movement and change to the undifferentiated flow of life? In other words, even if we were to repeat after Spinoza and then Deleuze that "we do not know what a body can do"[22] and thus embrace the unacknowledged potential of human and nonhuman entities to transmute and produce things, perhaps *knowing the difference* between different acts, passions, and forms of knowledge is more significant for understanding creativity and for thinking creative media than dwelling on the yet-unrealized and yet-unknown corporeal potential. (Incidentally, this act of knowing the difference, of being able to differentiate between positions and affects, is what the Kantian tradition calls "critique," a point to which we shall return in a short while.)

Yet post-Bergson, and especially post-Deleuze, it seems difficult for many theorists to talk about this movement and creative expansion of life "neutrally," so to speak, without falling into rhetorical hubris. In many works inspired by vitalist philosophy, life tends to be positioned as a secular miracle, a sublime force that evokes both admiration and awe. So when the philosopher Roberto Esposito, in his reading of Nietzsche, describes life as not knowing "modes of being apart from those of its own continual strengthening"[23] and as being "ecstatically full of itself,"[24] he changes, with Nietzsche's help, the "negative" Darwinian (if not necessarily Darwin's own) valorization of life as bearing an "original deficit" to its positive evaluation in terms of "exuberance and prodigality."[25] Similar effusiveness can be found in the writings of many others influenced by the philosophy of immanence to be found in "life," in which theoretical excursions into the natural sciences and, in particular, molecular biology, often result in the rather gushing descriptions of life's force, energy, and proclivity, with "life" redressed as the eighteenth-century Casanova on the way to his desirous conquests. To offer another example, we can look at John Protevi's formulation of "joyous affect" as a description of a preference in Deleuze, a preference which for Protevi indicates a normative dimension in Deleuze's philosophy.[26] In his reading of Protevi, Jeremy Gilbert explains that "joy is that which enhances the 'powers of life,' and as such is always understood in terms of becoming: mutability, movement and change, while any analytical assumptions as to the fixity of given categories or identity must always be overcome in the pursuit of it."[27] This preference for "life/becoming" over "stasis/being" underpins many contemporary philosophies of life that have made serious inroads into discussions of politics, culture, and art.

But what does it really mean to say that life is all-encompassing, that it is overcoming itself and pushing forward? Are all forms of life the same, and do they all have the same effectivities? Does this originary underlying layer called "life" actually "exist"? Or is this postulated unity of the plethora of natural variations and transformations constantly occurring in various organisms itself a fantasy, a projection, and a project-in-the-making? In attempting to think creativity and emergence in the universe at different cosmic levels by drawing on genetic variation, many theorists seem to have posited life as both an idea and ideal that has subsequently been used to provide a foundation for their particular philosophical framework (i.e., affirmative vitalist philosophy)—a framework that requires this postulated multiplicitous unity as its condition.

In our attempt to "find the conditions under which something new is produced (*creativeness*)"[28] we do take inspiration from Deleuze and, in particular, from his reading in *Bergsonism* of life as "production, creation of differences."[29] However, we also want to raise some questions for such a positive valorization of the creative force of life as a good in itself and as an adequate foundation for the philosophy of creativity. The redemption of desire by Deleuze and Guattari from its psychoanalytic associations with constraint and lack, and its reinscription as a positive force that is capable of enhancing the power of bodies and establishing connections between them, seems to entail a blind spot: the ascription to life's genetic push to expand and grow a *value judgment* and *a meaning* (via Bergson and Nietzsche), and hence an interpretation that is otherwise decried in the Deleuzian method. Undoubtedly, the adoption of at least some minimal starting conditions—such as life as a creative nonpersonal force—is necessary for any philosophy. What is more, the authors of *Anti-Oedipus* and *A Thousand Plateaus* are clearly aware of the "fabulatory function"[30] of their own work that manufactures conceptual and figurative giants that then gain a life of their own. With this in mind, it is the rather brusque and literal adoption of Deleuze and Guattari's vitalist formulations by some media scholars—an approach we could describe as a defabulation of Deleuze and Guattari's philosophy—that troubles us here, especially when it takes the form of a belief in the unproblematic inventiveness of desire as a creative force. As Gary Hall and coauthors argue,

> Many of those writing under the influence of the philosophy of Gilles Deleuze and Félix Guattari have interpreted desire somewhat crudely--as an inherently positive, primal propulsive force that opposes domination and control. Within the realms of critical and cultural theory, in particular, this has indeed led to a marked shift in emphasis and attention: away from a concern with representation and critical interpretation, and toward the generation of creative proliferations of desire.[31]

Making *Sense* of Vitalism

In his *Parables for the Virtual*, Brian Massumi starts from an engagement with Bergson's notion of duration and explores the consequences of looking at movement as primary and the emergence of positionality and stasis as secondary. This examination leads him to propose a new framework for cultural engagement, one that does not "elide nature"[32] and that remains attentive to the becoming of matter. But then Massumi's argument takes a rather unexpected turn. He goes on to suggest that "if you want to adopt a productivist approach, the techniques of critical thinking prized by the humanities are of limited value."[33] This conclusion is justified by his argument that critical thinking is not a discrete activity that can distantly look at an object but that it always also produces things, as any "activities dedicated to thought and writing" are inventive, in the sense that they feed back and thus contribute to what we know as "reality." So far, so unproblematic (for us, at least). Yet Massumi's call for academics to "shift to *affirmative* methods: techniques which embrace their own inventiveness and are not afraid to own up to the fact that they add (if so meagerly) to reality"[34] seems to be premised on the erasure or forgetting of criticality, which ultimately becomes a stumbling block in his otherwise exciting proposition. The sheer recognition of the fact that analytical work is never just that, or the valorization of one's academic work as work of invention—say, that of a cultural sociologist who is analyzing the creative industries but who is also aware that she is actively shaping the reality of what is being referred to as "cultural industries," of the discipline of sociology, and of herself as a cultural subject—needs a return to criticality for that scholar to be able *to make this realization* and to differentiate it from other types of academic research. It is clearly not enough merely to expect her to "just do it" and to continue researching through growing, expanding, and flourishing—the way, say, yeast does?[35]

This point leads us to another question. Even if we do take up Massumi's challenge to recognize academic work as always already the practice of invention and follow his call to shift from critical to affirmative methods, how can we *make sense* of what is being created and of our own creations? In other words, how will we ensure that whatever emerges through our creative process is worth our time, how will we tell whether it is progressive or conservative? (Unless we are to be satisfied with *any* change, *any* transformation, just for the sake of it.) Such affirmative and creative processes to be embarked upon, or joined, by academics and artists do not of course appear in some kind of nebulous cosmic soup that has the same viscosity and temperature for everyone everywhere; rather, they take place in the discrete set of circumstances, stabilizations, or what Massumi calls "positionalities" and "stases." And then some of these positionalities and stases matter more than others; they all have different affordances and different consequences, all of which will need to be *critically thought about* at some point. Surely it is not enough to rely only on desire—even if understood as a nonhuman force and "the condition of evolution"[36]—as a primary force behind such inventiveness?

Again, even if we take desire as the intention-free unfolding of matter through time, it is impossible to separate this understanding from its particular, temporary human incarnation, in the sense that what a patch of mold growing in our fridge "desires" is most likely rather different from our own desire, just as what we desired yesterday may be rather different from the desire that will drive us tomorrow. Perhaps if we stick with the more vitalist understanding of desire as a push for life to expand, we could impose such a parallelism, which would also liberate us from the humanism of interpreting our own actions as being regulated primarily by volition. But then, even on Massumi's own terms, we cannot do just this, as he mobilizes the notion of desire to talk about the unfolding of Stelarc's artistic practice over time—a rather singular set of artistic interventions that may also include the element of the body evolving over time with and as machine, beyond the control of the artist, but that is not free from the artist's own interventions and judgments. Consequently, even if we were to concede that desire is both "this" and "that," we have to agree with Hall and colleagues when they say that "not all forms of desire can be regarded as being psychologically or politically progressive. Nor do all instances of desire necessarily oppose domination. Some instances of the desiring field can be neo-liberal, totalitarian, even fascist."[37] This leads Hall and colleagues to ask the following, rather significant, questions: "How do we *judge* which creative proliferations of desire are politically just and progressive, and thus capable of producing deterritorializing 'unblockages,' psychologically and socially? Who *decides* which desires oppose domination and offer escape plans from already mapped out existential and philosophical paths and which do not? On what basis, on what grounds, can such judgments be made?"[38] In a piece that provides an introduction to their film essay on Deleuze's "Postscript on the Societies of Control," Hall and coauthors are not writing *against Deleuze*, only against what they are referring to as rather uncritical (in every sense of the word) "Deleuzianism," also termed—after the Italian writer Bifo—the "desiring movement."[39]

From Vitalism (Back) to Critique

We therefore want to propose that an affirmative vitalist philosophy needs to be exposed to, and engage with, a critique, unless it is prepared to disappear down its own metaphysical blind alleys—although rebranded as nature's corridors. So, far from seeing critique as an unwelcome relic of certain types of academic practices before "the affective turn,"[40] we are positioning it as a precondition of academic and artistic creativity. Criticality can save us from what we are terming "creative mania," a desire-driven chase for originality that naïvely replicates the very structures and strictures of Romantic creation—albeit now dressed in the language of materialist-vitalist philosophy, with some sprinkling of biology. And thus, when, for example, Matthew Fuller suggests in his book *Media Ecologies* that media technologies need to be understood as assemblages rather than wholes and that technology itself needs to be seen as a bearer of forces and drives, he is making an important intervention into the divisive—or what Bergson would call "solidified"—readings of machines that situate agency on the side of the human but that do not seriously think about humans' coemergence with and through *tekhnē* (an argument that forms the core of Bernard Stiegler's *Technics and Time*, as well as many of Gilbert Simondon's and Deleuze and Guattari's writings). But what is missing from Fuller's discourse is a further discussion of the premises behind, and consequences of, such connectivity—presumably because that would take us too far down the "analysis" path. What we get instead is an incessant reiteration of this "connectivity" and "relationality," which, through the rhetorical force of his argument, is positioned as fact. There is no closer look at what he calls the "minor processes of power" at work in pirate radio stations, mobiles phones, or on the Internet, only jumps across molecular levels, sliding up and down the scale of the universe in an attempt to claim this intermeshed "media ecology." Given that, like many other "molecular theorists," Fuller does not pay particular attention to language, treating it as a semitransparent tool that allows him to describe the networked connectivity of the world, its evident figurativeness becomes something of an obstruction in his study of the "machinic phylum."

Our own understanding of critique is partly derived from the work of Foucault, although we also want to raise questions for some of its human—all too human—delimitations. What is most interesting about Foucault's use of the term "critique" for us is that it goes beyond its Kantian conceptualization in terms of recognizing the limits of knowledge and going against those limits. For Foucault, critique consists of putting constraints on, or even refusing, authority—of the Church, of scripture, of the law, and governance. It is "the art of voluntary inservitude, of reflective indocility,"[41] which is focused on transforming the relations between power, subjectivity, and truth.[42] The notion of truth is vital for Foucault, but what he has in mind here is a "particular truth" to be created by each self as a relationship to what is not in it, not a preexistent general truth already in place, awaiting discovery. The self needs to establish a connection to a truth outside of itself, which it has to work out, memorize, and progressively put into practice. This idea of a particular truth to be arrived at and a particular way of life to be created foregrounds one of the most significant aspects of Foucault's nonprescriptive normativity: the absence from it of any prior codification or instructions on how to live. It is the ethical activity itself—whereby ethics stands for a practice of freedom derived from the game of truth—that is important here, rather than the fulfillment of any particular commandments.[43] Foucault's insight that "critique only exists in relation with something other than itself"[44] may sound obvious, yet it attaches this practice to singular objects and anchors it in singular events and instantiations.[45]

Judith Butler argues that critique "not only suspends judgment for [Foucault], but offers a new practice of values based on that very suspension."[46] From this, critique presents itself to us as more than "fault-finding": it is first and foremost a creative practice that *temporarily* suspends judgment but that ultimately requires a decision. If, according to Foucault, "There is something in critique that is related to virtue,"[47] then critique also has a normative character. It enables making distinctions between states of being and encourages envisaging new, better states of being. In the words of Isabell Lorey, critique has a potential to initiate "a recomposition," "the constitution that emerges and results from non-standardized heterogeneous practices, from practices that dare to invent something new."[48] More than a theory of judgment or a critical analysis of categories, critique can thus perhaps be understood as a theory of invention, even a theory of "inventing well." We can see from this that any suspension of judgment can only ever be temporary and that if we do not want to concede that any form of novelty and invention is good per se (i.e., if we are not prepared to read any new form of mutation of life or technological invention as a good in itself in advance—which we are not, in this book), then we need to bring back a theory of decision of some kind.

Drawing on Derrida's understanding of the relation between politics and undecidability, Hall and coauthors claim that any decision with regard to a specific instance of media activism or creative proliferation of desire "*cannot* remain entirely open and incalculable." Instead, they argue,

> it has to be based on rationally calculated and reflected upon—in however compromised and limited a fashion—*universal* values of infinite justice and responsibility. At the same time, any such decision cannot be made solely on the basis of knowledge and values that have been extensively thought about and decided upon *a priori*, such as a preconceived political agenda or theory (and that would include any preconceived position derived from Deleuze and Guattari's schizo-analysis). To make a responsible political decision it is necessary to respect *both* poles of the non-oppositional relation between the calculable and the incalculable, the knowable and the unknowable, the decidable and the undecidable; but then to make a leap of faith.[49]

Such a leap of faith in the face of a "certain non-knowledge" which leaves us "disarmed," but to which we feel "freely obligated and bound to respond,"[50] can perhaps serve as another name for a creative process that also remains critical, which we are trying to envisage here. But whom will these leaps of faith be made by? How can we account for the critical agency of that decision maker who temporarily cuts through this process of constant becoming in order to make a mark on life, and for the context in which such an event takes place?

Critical Attention

In his reading of Foucault, Gerard Raunig attempts to enact a synergic merger between the two senses of critique in Foucault as "suspension of judgment" and "reinvention" when he says that "what Marx and Engels called 'practical critical action' is already quite close to the Foucaultian concept of the 'critical attitude' in this emphasis on turning away from Kant's purely epistemological critical project."[51] This form of critique is thus always already "practical" and "political" in that it enables the transformation of not only the self and its relation to truth, but also of the sociopolitical and material context in which the self operates. We are thus now faced with the need to differentiate between judgment and decision and also to consider the creative force and possible limitation of subjectivity in starting off this kind of transformative process. Indeed, in Foucault's critical attitude it is still very much the (human) self that is the pivotal point of the transformation: the self's agency serves as a yardstick for evaluating change in the outside. Ewa Ziarek acknowledges the limitations of Foucault's reaching to the outside, whose ontology always seems to return to the self as its starting point, yet she also considers the possible overcoming of that agency through the event itself once it has been set in action. Quoting Foucault's statement that "no one is responsible for an emergence; no one can glory in it,"[52] she suggests that agency needs to be rethought beyond its humanist conceptualization as a "different configuration of forces."[53] The event that ensues, and that marks a rupture within the status quo, for her "exceeds not only the current configuration of power but also the intentionality of the subject."[54] It is precisely this tension that facilitates "the creation of the new forms of life—new modalities of pleasure, new modes of relations, new ways of being,"[55] with an ethical dimension of his critique being turned toward the invention of new, less constraining forms of life.

Drawing on Foucault's notion of "critical attitude," we want to propose a somewhat modulated term, "critical attention," which stands not only for an ethical opening and an injunction to both receive well and produce well but also for a mindful and corporeal disposition. It is a concept that makes more use of this entanglement between us and the world, where, in Karen Barad's words, "To be entangled is not simply to be intertwined with another, as in the joining of separate entities, but to lack an independent, self-contained existence."[56] For Barad, existence is "not an individual affair," which is another way of saying that time and space, matter and meaning, "are iteratively reconfigured through each intra-action, thereby making it impossible to differentiate in any absolute sense between creation and renewal."[57] "Critical attention" thus transcends human-centered intentionality by foregrounding the "entangled state of agencies"[58] at work in any event. This is not to backtrack on what we earlier described as the theory of an inevitable but also somewhat impossible decision at the heart of which always lies a leap of faith, or to deny singularly human, ethicopolitical responsibility for such events. But it is certainly to acknowledge that what we are referring to as "human" is only a distinct entity "in a relational, not an absolute, sense" because, as Barad explains, "*agencies are only distinct in relation to their mutual entanglements; they don't exist as individual elements.*"[59]

Incisions as Decisions

Any emergence of creative media will therefore always be a coemergence in which *we* shall also take part and through which we shall also become "something else." Yet even if we *are* hinting at the mutual co-constitution of the human and the world, and the entanglement of the processes of natural and technical evolution, we are not suggesting here that media can grow like yeast, bacteria, or babies and hence imposing an unproblematic parallelism between biological growth and cultural production. "Critical attention" as a material-philosophical opening toward the emergence of the new is one way of guaranteeing that "we" will be able to both tell the difference and make a difference.

To sum up what we have said so far, though it remains in conversation with many theories and positions that can be broadly described as "Deleuzian," our proposal for taking, after Bergson, creativity as the creation of forms ever new is nevertheless premised on some modulated assumptions and postulates:

1. Life is a neutral force of expansion and movement that "just is" or rather "*just is becoming*"; it is a minimal condition of being in the world, and of there being a world, but not necessarily a good in itself.
2. This neutrality is only ever a precultural and somewhat fictitious point of origin, as in specific theological and philosophical frameworks, and in specific cultural circumstances, different processes and forms of life may be valued as more or less desirable.
3. Following Bergson, analysis can be seen as a limited (even if at times desired or needed) form of enquiry, as it requires us to ignore the creative movement of life, and the perpetual changes occurring within it,[60] as well as our own coemergence with many of the entities that become objects of our analysis.
4. A Bergson-inspired intuitive method of knowing, which "places itself in mobility, or . . . duration,"[61] allows for the recognition and generation of new forms of life—and new forms *in* life—beyond the replication of what is already there, or what Bergson calls "solids."

5. What Bergson calls "solids" and what we term "temporary stabilizations of life" exert a significant and significatory cultural and material power and demand; they cannot be overcome once and for all in any straightforward manner, but must instead by attended to, always in a singular way.
6. "Cuts" made into duration have not only ontological but also politicoethical consequences, which require critical attention.

Stripped of its earlier associations with divine creation, creativity understood as an expansive force of life can therefore perhaps be positioned as a minimal condition for engaging with the world and its processes and entities. Another way of saying this is just consenting that "life is," or even that "life is (a) becoming," without drawing any particular conclusions or applying any a priori valorizations to this fact. This consent involves withholding any judgment on the positivity of its accumulation or unfolding, and the possibility of agreeing that life "as such" may not "exist"—although of course in particular circumstances and with relation to particular life processes and living beings (including most often, although not always, our own being), we may have to decide that such a becoming is desirable and that its diversification should be encouraged. However, it is not from within life "as such" that such a decision can ever be made: it can only come from *bios*, that is, the organized life of a human who finds him- or herself in the position of being able to ask such questions and make such valorizations, unstable, compromised, and temporary as they may be. Naturally, sometimes—most often, perhaps—such decisions with regard to the expansion of life or its contraction will indeed be taken by "life itself," by viruses, bacteria, cells. Yet these can be described as "decisions" only in a metaphorical sense and need to be differentiated from the perhaps rather rare but not any less significant instances of human decisions about life taken in an undecidable terrain.[62]

There Is No Creation without Experiment[63]

The very process of writing *Life After New Media* has been for us an attempt to let such critical creativity unfold and to experiment with a slightly different mode of doing critical work on media, beyond what Deleuze has termed, after Foucault, "the author-function."[64] And thus, even though it takes the form of a fairly conventional academic monograph, this book is also a kind of "live essay," performed in (at least) two voices via numerous exchanges of electronic traces, graphic marks, face-to-face utterances, and corporeal gasps. Writing academic scholarship in the form of such a live essay aims to facilitate collaborative thinking and dialogic engagement with ideas, concepts, and material objects at hand between the book's authors, or rather conversational partners. There is of course a tradition of such endeavors in the humanities: we can think here of Hélène Cixous and Catherine Clément's *The Newly Born Woman*, or of Gilles Deleuze and Félix Guattari's joint works—in the writing of which "there was already quite a crowd," with each of the authors identifying themselves as "several."[65] Yet we are also aware of the inevitable humanist (re)turn of any such collaborative effort, with temporary freezings, disjointed positions, and singular inventive "cuts" disrupting the discursive emergence of the text and project at hand.

One of the reasons for our interest in developing such a critical yet creative media work is our shared attempt to work through and reconcile, in a manner that would be satisfactory on both an intellectual and aesthetic level, our academic writing and our "creative practice" (fiction and photography, respectively). This effort has to do with more than just the usual anxieties associated with attempting to breach the "theory-practice" divide and trying to negotiate the associated issues of rigor, skill, technical competence, and aesthetic judgment that any joint theory-practice initiative inevitably brings up. Working in and with media is for us first and foremost an epistemological question of how we can perform knowledge differently through a set of practices that also "produce things." The nature of these "things"—academic monographs, novels, photographs, video clips—is perhaps less significant (even though each one of these objects does matter in a distinctly singular way) than the overall process of producing "knowledge as things." In other words, what we term here "creative media" is for us a way of enacting knowledge about and of the media by creating conditions for the emergence of such media. We envisage such "creative media" works to be situated across the conventional boundaries of theory and practice, art and activism, social sciences and the humanities. They can take a variety of forms—essays on, polemics with regard to, and performances of what it means to "do media" both creatively and critically. They can also incorporate a variety of media, from moving and still images, through to interactive installations, codework, creative writing, and more traditional papers. (And yes, language also counts as a medium.) We recognize that there may be something rather difficult and hence also frustrating about this self-reflexive process, whereby it is supposed to produce the thing of which it speaks (creative media) while drawing on this very thing (creative media) as its source of inspiration—or, to put it in cybernetic terms, feedback.

Yet this circularity is precisely what is most exciting for us about the theory of performativity and the way it has made inroads into the arts and humanities over the last two decades. Drawing on the concept of performativity taken from J. L. Austin's speech-act theory as outlined in his *How to Do Things with Words*, thinkers such as Jacques Derrida and Judith Butler have extended the use of the term from being limited to exceptional phrases that create an effect of which they speak (such as "I name this ship *Queen Elizabeth*" or "I take this woman to be my lawful wedded wife") to encapsulating the totality of discourse (which in itself is never total or complete).[66] In other words, any bit of language, any code, or any set of meaningful practices has the potential to enact effects in the world, something Butler has illustrated with her discussion of the fossilization of gender roles and positions through their repeated and closely monitored performance. Performativity is an empowering concept, politically and artistically, because it not only explains how norms take place but also shows that change and invention are always possible. "Performative repetitions with a difference" enable a gradual shift within ideas, practices, and values even when we are functioning within the most constraining and oppressive sociocultural formations (we can cite the Stonewall riots of 1969, the emergence of the discipline of performing arts, or the birth of the Solidarity movement in Poland in 1980 as examples of such performative inventions). With this project, we are thus hoping to stage a new paradigm not only for *doing media critique-as-media analysis* but also for *inventing (new) media*.

The Invention of What? (And What For?)

Of course, not all events are equal, and not everything that "emerges" is good, creative, or even necessarily interesting. Far from it. Mediation, even if it is not owned, dominated, or determined economically, is heavily influenced by economic forces and interests. This state of events has resulted in the degree of standardization and homogenization that we continue to see across the board: witness the regular "inventions" of new cell phones or new forms of aesthetic surgery. The marketization of creativity ends up with more and more (choice) of the same—even if some of these "inventions of the old" can at times perhaps be put to singularly transformative uses. Yet most events and inventions are rather conservative or even predictable; they represent theater-as-we-know-it. Our own investment lies in recognizing and promoting "theater-as-it-could-be." The latter phrase is adapted from Chris Langton's founding definition of artificial life—a field we looked at in chapter 5 that is underpinned by the most conventional metaphysical assumptions about science and life, yet that manages to open a network of entirely unpredictable possibilities for imagining "an otherwise world."[67] We are interested in witnessing or even enacting the creative diversification of events as a form of political intervention against this proliferation of difference-as-sameness.

We find such "noncreative" diversification everywhere, including in the increasingly market-driven higher education, which aims to prejudge and quantify scholars' and artists' "impact" on economy and society. One can easily blame "performance audits" such as the Quality Assurance Agency's inspection visits and the Research Assessment Exercise or the new Research Excellence Framework in the UK or the compiling of international university league tables for the standardization and homogenization of the academic output worldwide. But these "quality-enhancement" procedures are just a means to an end—this end being competition and survival within an overcrowded global market, run on an apparently Darwinian basis whereby size (of institution) and volume (of output) really do matter. In this kind of environment, it is sometimes very difficult to make a difference.

But we can remind ourselves here of Donna Haraway's willingness to recognize the real limitations of a set of practices she referred to as "cyborg politics." In that old, seemingly dated "battle of the cyborgs," she was always going to lose, but never to concede: the latter would have been like leaving "in the hands of hostile social formations the tools that we need to reinvent our lives."[68] Haraway's cyborg politics may have been a feature of the Cold War, but what survives of it is the politicoethical injunction to intervene, to make a difference, not for the sake of difference itself, but for the sake of a better—more interesting, more just—world. We take the "making" in making a difference as seriously as the difference itself. Hence our insistence that theory takes the form of theater, that it is always already performative, and hence our quest for the "invention of forms ever new"—to use Bergson's term. These forms are hybrid, recombinant—and challenging. They represent the kind of conceptual risk taking and creativity that Rosi Braidotti calls for in her book *Metamorphoses* and that emerges from feminist philosophy, among other places. There comes a point, Braidotti insists, when it is no longer enough to deal with the breakdown of hierarchical conceptual dualisms just in the content but not in the form of our address. As soon as we attempt to performatively engage form and content, reason and imagination, we are faced with the controversial question of style that relates to the academic conventions of argument and presentation. Hence we are more than willing to join Braidotti when she says, "I do not support an assumption of the critical thinker as judge, moral arbiter or high-priestess."[69] In consequence, an alteration in the traditional pact between the writer and her readers inevitably takes place: the "writer/reader binary couple is recombined," Braidotti says, "and a new impersonal mode is required as a way of doing philosophy."[70]

"They Branch Out and Do Not Stop Branching Out"[71]

One of us (Sarah Kember) has been pursuing such an "impersonal mode" in her attempts to write science and fiction in a way that fully recognizes their mutual (re) mediation, as well as their existence as relational, non-self-identical *differences in kind*. Traditional literary fiction can be said to "other" science by either avoiding it or subsuming it within familiar humanist narratives. Science fiction, in turn, tends to fetishize (or demonize) what literary fiction elides. In so far as there are processes of othering at work in the attitude of science to fiction and, to an extent, of fiction to science, then what exactly lies between them? The impersonal mode requires experimentation in form and content and entails technical difficulties and problem-solving abilities on a surprising scale.[72] In terms of writing, the chief among these is precisely *how* to reconcile the exterior and interior world views normally associated with the sciences and arts, respectively. Traditionally, fiction offers a view of the world from the inside out, while scientific and academic writing offer a view of the world from the outside in. In her experiments with writing across these two worldviews, Kember is learning, the hard way, what is involved in crossing this tradition.

Her *The Optical Effects of Lightning* is a story, possibly a true one, about an experiment in human cloning and what it means to two narrators who are twins (clones may be thought of as twins, separated in time). One narrator speaks in what Braidotti might call the "judgmental, moralizing high-tone" of someone who is located or considers himself to be located outside the experiment—commenting and reporting on it and on the protagonists involved. The other narrator speaks of his involvement in the experiment, and of his experience, in a more conventionally fictional or interior voice, which becomes increasingly obsessive and perhaps deluded. His less than likely, incredible, highly subjective narrative is framed by the other, more controlled, more rational and objective one, but we are never entirely sure which one is "true." The story complicates this problem of truth when, like two cells in a cloning experiment, the narratives and their narrators are literally fused. This attempt at literary science fiction plays on an analogy between the electrofusion of cells and the optical effects of lightning on the body.[73] The intention, or rather intervention, here is to assert that scientific processes such as fusion are more than metaphors for fiction, or indeed, for theory. Instead, they become materialized and play a performative role *within theory*. This sort of intervention is therefore also something of an invention.

To a certain extent, this experiment is reiterated in Kember's *Media, Mars and Metamorphosis*. Here, the narrator, Jeremy Hoyle, is an academic and cultural commentator; he is a Fukuyama-like figure[74] concerned with three life-changing experiments in biotechnology. These experiments relate to different spatial realms, but are linked, in part, by their focus on the cell. They incorporate outer or cosmic space, the interior space of the computer, and bodily space at the boundary between self and other. These different spaces thereby become analogous. The experiments—in bacteriology, immunology, and mediology—include one to test for the presence of microbial life on Mars, another designed to induce tolerance in face transplant surgery, and a third, user-based experiment to test for prospects of intelligent media. Three different characters—Lou, Hannah, and Hal—talk about the life-changing experience of being involved in these experiments. Lou, an elderly and embittered microbiologist, whose previous claims to have discovered evidence of life on Mars have been repeatedly rejected by NASA, declares finally to have evidence of a Martian microbe with characteristics similar to that of green sulfur bacteria. Hannah, a neurotic young woman involved in a traumatic act of violence, claims to have had the first successful face transplant, based not on immunosuppressant drugs but on the establishment of immune tolerance and hybridity between the donor and recipient. Finally, Hal, a middle-aged curmudgeonly technophobe with a drink habit who agreed to take part in a smart home experiment because he needed the money, claims that something "weird" happened when the speech-based, adaptive, and so-called intelligent objects he was forced to interact with started to sound more and more like him. Jeremy interviews each character and—understandably—does not really believe them. Not only do these experiments, and others like them, cross the line of good science and the sanctity of human nature, but they are also more than likely to be hoaxes. However, his conclusion is somewhat complicated by his own subsequent experience. Jeremy starts to feel ill. He is not himself. He must have a terrible stomach bug or something because he is shitting green stuff and hallucinating. He does not even recognize his own face in the mirror. What's more, and what's worse, neither does the mirror. And it's telling him so.

The point of these true stories—these "factions" that stay as close as possible to what is happening in the world of technoscience now—is not to validate the humanist category of experience but rather to explore the possibilities of what Keith Ansell Pearson terms "experience enlarged and gone beyond."[75] That is to say, experience gone beyond anything singular and toward something potentially multiple and inherently nonexperiential. The body, poor Jeremy's body, enacts or performs this enlargement of experience—*for us*. We (writer and reader) attend his transformation, his metamorphosis, in as far as we identify with his rigid and righteous refusal of it. It is not a nice trick to pull on him, the author's substitute—the dramatized and somewhat parodied voice of the theorist. But perhaps it is time we dealt with our alter egos, canceled each other out as we are supposed to—at least in the Gothic literary tradition, if not in the academic one—and found a different mode of working and playing.

Technology and the Body

As mentioned earlier, Haraway is an important figure for us in our joint work and play because she was one of the first thinkers to offer a critical, insubordinate, and playful engagement with technological processes within a wider sociocultural setup. Even though her "cyborg," a Star Wars–era creature that hybridized flesh and metal, carbon and silicon, seems positively old-fashioned in the current era of biotechnological hybrids that can literally get under our skin (or into our digestive systems), the political significance of her intervention into what she termed "technoscience" has not lost any of its validity. For any creative media project to be truly inventive, it needs to work through the ontological and epistemological consequences of technologies and media becoming increasingly closer to us. It also needs to consider what the philosopher Bernard Stiegler calls our "originary technicity,"[76] in which technology is comprehended as an originary condition of our being in the world, not just an external object we all learn to manipulate for our advantage and benefit.

This is a very different view of technology and mediation from the one that sees the human as "natural" and technology and media as external agents. This view challenges the instrumental understanding of technology proposed by the Greek philosopher Aristotle, a framework that still shapes the majority of our media stories about IT, the Internet, or genetics. Within this instrumental framework, technology is seen as just a tool for the human. It is an external object that either promises us pleasure, if it is a gadget such as a digital camera, or threatens our life and well-being, if it is a bomb or a lethal injection. However, what we are trying to do with our alternative media paradigm is argue for the possibility of, and need for, adopting a different model—one proposed not only by Haraway and Stiegler but also by the Australian performance artist Stelarc. All these thinkers are very critical of the story of the human as a master of the universe who can become even more powerful via his media gadgets. Instead, they outline a more systemic and networked model of human-nonhuman relations, in which mediated prostheses are seen as *intrinsic parts* of the human body.

In an interview for Zylinska's edited collection *The Cyborg Experiments: The Extensions of the Body in the Media Age*, Stelarc explains his understanding of the relationship between technology and the human as follows:

> The body has always been a prosthetic body. Ever since we evolved as hominids and developed bipedal locomotion, two limbs became manipulators. We have become creatures that construct tools, artefacts and machines. We've always been augmented by our instruments, our technologies. Technology is what constructs our humanity; the trajectory of technology is what has

propelled human developments. I've never seen the body as purely biological, so to consider technology as a kind of alien other that happens upon us at the end of the millennium is rather simplistic.[77]

So, clearly, we should not think that there was once a "pure" body and that this has somehow been contaminated just as we entered the technological age. Instead, as Stelarc puts it, "We've been simultaneously zombies and cyborgs; we've never really had a mind of our own and we've never been purely biological entities."[78]

From this critical cybernetic perspective, the human is seen as having *always* been technological, or having *always* been mediated. To put it differently, technology and media are precisely what makes us human. Even if we agree that the body is somewhat weakened or inadequate in a world of ubiquitous information flows, computer-led wars, and nanotechnology, it does not mean we have to bemoan the loss of our human potency or desire to become Terminator-like robots ourselves. We can better understand this position as a pragmatic recognition of our dependency on technical and media objects. The work of technoartists such as Stelarc, or technophilosophers such as Stiegler and Haraway, should not therefore be reduced to a naïve prophecy of a postflesh world in which man will eventually overcome his technological limitations. Instead, we should rather see it as an exploration of the symbiotic relationship the human has always had with technology and media. In other words, it shows us technology as being an inseparable part of both "the human" and "the body."

Why is it important for us to think of ourselves in this way? To begin with, this position allows for a better understanding of what we have posited in this book as the relationality of the world—of which we are part. It also lets us develop a more interesting and more critical engagement with "nature" and "the environment." If we do accept that we have indeed always been cyborgs, that we have always been mediated, it will be easier for us to let go of paranoid narratives (such as Jeremy's from the above mentioned story) that see technology as an external other that threatens the human and needs to be stopped at all cost before a new mutant species—of replicants, robots, aliens—emerges to compete with humans and eventually to win the battle. All this is not to say that in the universe of complex relations between human and nonhuman beings "anything goes" and that all connections are equally good. But seeing ourselves as always already connected, as being part of the system—rather than as masters of the universe to which all beings are inferior—is an important step in developing a more critical and a more responsible relationship to the world, to what we call "man," "nature," and "technology." It is also a promise of the emergence of some more productive media relations and media environments.

Figure 7.1a–f
Joanna Zylinska, *We Have Always Been Digital* (2007), selected images.

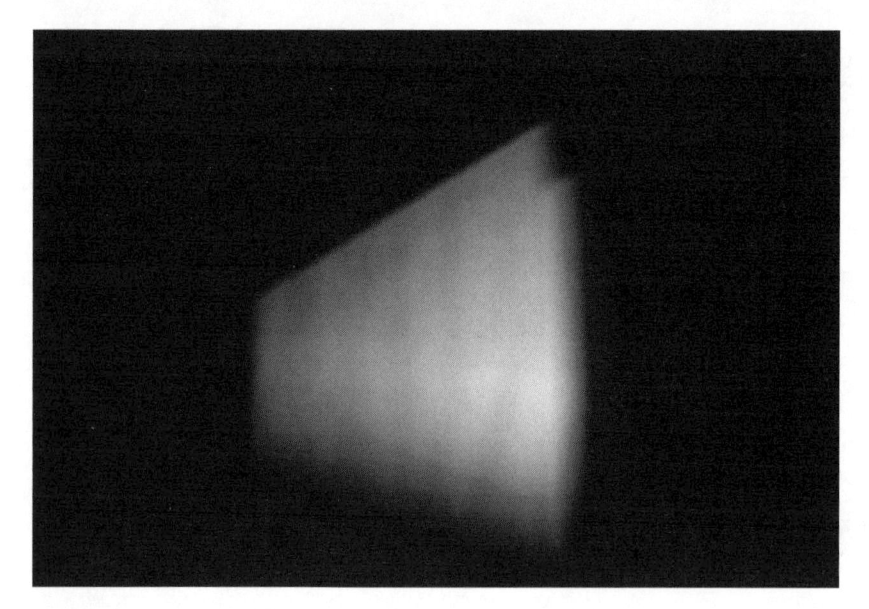

Figure 7.1a–f
(continued)

10001110101011000110011010011010001011011001011010101110
11000101011001001001001000100010010000010100111101110100
10010010101110010010010010001001001001001000010001111110
111100000000100010001000010001000100010001111100010000100
00111000010000111111010101010101010110000110011000111001
10010010000110001000111100101100111001001010100100010000
01001001110010101010101010110010101001101010001110101011
10011010001011011001011010101110001010101011000101011001
00100010010000010100111101110100111101000010010010101110
10001001001001001000010001111110100111111011111000000100
10001000100010001111100010000100011100111000111000010000
01010101010110000110011000111001001100100110010010000110
00101100111001001010100100010000100111001001001001110010
10110010101001101010001110101011000110011010011010001011
10101110001010101011000101011001001001001000100010010000
01110100111101000010010010101110010010010010001001001001
01111110100111110111110000000100010001000010001000100010
10000100011100111000111000010000111111010101010101010110
00111001001100100110010000110001000111100101100111001001

Figure 7.1a–f
(continued)

We Have Always Been Digital

Joanna Zylinska has attempted to enact such a productive relation between technology and the human in her photographic work as a way of actually *practicing media philosophy* and performing concepts via images. Her project *We Have Always Been Digital*, arose out of a desire to think about the media, both "new" and "old," and the way culture repurposes and remediates its different media forms. It may seem constraining or even reductive to begin discussing visual work with the verb "to think." Yet, to be interesting, creative practice, including photography, for us has to mobilize complex thought processes, although without doubt it should do *more* than just illustrate already worked-out ideas and concepts. The very nature of this "more" constitutes part of the invention process. Visual work can of course help us articulate concepts or states that exceed the linguistic and achieve things that spoken and written language cannot do. But then written texts themselves always already entail a certain lack of determination: even the tightest philosophical argument is always based on a leap of faith and remains underpinned by numerous investments of which we can be only partly aware. This is a roundabout way of saying that finding a satisfactory way of negotiating between visual and textual narratives is never going to be easy for a media theorist.

We Have Always Been Digital explores digitality as the intrinsic condition of photography, both in its past and present forms. Rather than focus on the aesthetic qualities of light, it invites the viewer to consider the formal role of light in the constitution of a pattern, the "ON/OFF" of the information culture. The project assumes that computation also takes place outside what we conventionally think of as "computers." Indeed, it is through the differential effect of the presence and absence of any data—of pattern, electricity, light—that computation occurs in the wider world, engendering complexity and bringing about change. The images show the digital flow and exchange of data in different media: house walls, furniture, human bodies. They capture the digital condition: the emergence of a pattern of zeros and ones.

The project has numerous affinities, both textual and visual. It starts as a remediation of W. H. Fox Talbot's "photogenic drawings" of lace and of light falling through the window panels in Lacock Abbey—a set of images Talbot allegedly sent to his friend Charles Babbage, the inventor of the differential engine (the first computer). This story, hinting at the parallel invention of photography and computing as two ways of capturing the pattern in different media, has been narrated by Geoffrey Batchen in his article "Electricity Made Visible." For Batchen, "photography is a binary (and therefore numerical) system of representation involving the transmutation of luminous information into on/off tonal patterns made visible by light-sensitive chemistry."[79] It is therefore a fledgling form of informational culture, one that since its inception records the presence and absence of data. The images that constitute *We Have Always Been Digital* have also been infused with some splinters of the theory of computational universe developed by the likes of Edward Fredkin, which assumes that all matter is computational, that is, that it consists in the differentiation between present and absent bits of information.[80] However, any straightforward applications of this theory as allegedly telling us "what the world is like" are bound to be deeply problematic. The singular materiality of each photographic medium—be it camera, paper, computer screen, or human body, from which the image is emitted and on which it is projected—destabilizes the universalizing seamlessness of Fredkin's propositions. It is precisely in the tension between the conceptual and technical "liquidity" of the photographic object *as such*, and the (always temporary) solidity of its medium, that interesting creative possibilities are arguably opened up. The point of such creative experimentation with the photographic medium lies not so much in nostalgically harking back to older, more "solid" media—be it analog film or mechanical large-format cameras—although sometimes these particular choices may present themselves to us as aesthetically better and more fitting. What is much more important, however, is not foreclosing the performative invention of photography, or any other medium—an invention that is always potentially ongoing, even if not always enacted—with ready-made decisions about its effects and affects, aesthetic, ethical, or political ones.

Ethical In-Difference

Even though human agency does not withdraw altogether from these processes of creative and critical invention, it is distributed throughout a system of forces, institutions, bodies, and nodal points. This acknowledgment of agential distribution—a paradox that requires a temporarily stabilized self which is to undertake this realization—allows for an enactment of a more hospitable and more enmeshed relationship with technology and the media. This notion brings us to an interesting point made by Stelarc, who, when commenting on his performances, mentioned adopting "the posture of indifference" in relation to them.[81] Such a posture involves abandoning any desire to entirely control the event and allowing it to just unfold after starting it off. Stelarc's pronouncement conveys a tension between the modernist notion of artist as lone creator and instigator of ideas on the one hand and the cybernetics-informed understanding of artist as a node in the network of exchange on the other.

Naturally, the decision about adopting this posture of indifference—about not having any expectations—is made by him, from a temporarily stabilized point of human agency. Still, we should perhaps read it as not *just* a rational decision but also as bodily passivity, as letting oneself be-together-with-difference, with-technology. To cite Butler again, "At stake here is the relation between the limits of ontology and epistemology, the link between the limits of what I might become and the limits of what I might risk knowing."[82] It is via points of temporary stabilization between human, corporeal, and technical agency that partial decisions are being made, connections between bodies are being established, aesthetic and political transformation is being achieved, and power is taking effect over different parts of "the network" in a differential manner.

There is no guarantee that temporary outcomes of any such ongoing performative processes will be critical and transformative rather than just repetitive and conformist. But it is this possibility of emergence of such transformations and inventions, of making *a difference that matters*, that can turn this creative media project from the theater of mere form to an ethicopolitical performance.[83]

Conclusion: Creative Media Manifesto

As well as offering an overview of the book's argument, our conclusion is designed as a double call: to arms and to arts. This call is aimed at intellectuals, writers, philosophers, artists, analysts, scientists, journalists, and media professionals who are prepared to say something about the media that extends beyond the conventional forms of media analysis. With this, we want to put our argument on mediation as the dynamic logic of the media to a more creative use, so to speak, in order to start thinking about "doing media studies" otherwise. Taking seriously both the philosophical legacy of what the Kantian and Foucauldian tradition calls "critique" (as discussed in chapter 7) and the transformative and interventionist energy of the creative arts, we are interested in envisaging and enacting playful, experimental, yet rigorous cross-disciplinary interventions and inventions that will be equally at home with critical theory and media practice and that will be prepared and able to make a difference—academically, institutionally, politically, ethically, and aesthetically.

Our common conjuncture in the discussion lies with what we are calling a "creative media project." The project arises out of our shared dissatisfaction with many ways in which media are currently looked at within the nexus of academic disciplines that have commonly become known as media, communications, and cultural studies. In their more orthodox incarnations as developed from sociology, politics, and communications theory, media, communications, and cultural studies typically offer analyses of media as objects "out there"—radio, TV, the Internet. Mobilizing the serious scientific apparatus of "qualitative and quantitative methodologies," they study the social, political and economic impact of these objects on allegedly separable entities such as "society," "the individual," and, more recently, "the globalized world." What is arguably lacking from many such analyses is a second-level reflection on the complex processes of mediation that are instantiated as soon as the media scholar begins to think about conducting an analysis—and long before switching on a TV or an iPad.

Through instantiating this project in a "manifesto" style, we are making a claim for the status of theory as theater (there is an etymological link between the two, as Jackie Orr points out),[1] or for the performativity of all theory—in media, arts and sciences; in written and spoken forms. We are also highlighting the ongoing possibilities of remediation across all media and all forms of communication. From this perspective, theater does not take place—and never did—only "at the theater," just as literature was never confined just to the book, or the pursuit of knowledge to the academy. What is particularly intriguing for us at the current media conjuncture is the ever-increasing possibility for the arts and sciences to perform each other, more often than not in different media contexts. Witness the theater that involved the mediation of the Big Technoscience project in September 2008: the experiment with the Large Hadron Collider at CERN, which we discussed at length in chapter 2. As explained there, the LHC is a particle accelerator used by physicists to study the smallest existing particles, and it promises to "revolutionise our understanding, from the minuscule world deep within atoms to the vastness of the Universe" via the recreation of the conditions "just after the Big Bang."[2] Rarely since the Greeks has such an attempt to stage metaphysics been undertaken with an equal amount of pathos and comedy—and it's now done with satellite TV networks staging the event for the worldwide audiences in real time! It is thus unsurprising that Lisa Gitelman should position media as "the scientific instruments of a society at large."[3] Performances such as the LHC often incorporate their own critiques, although these critiques tend to remain latent or unacknowledged. By remediating such events via critical and creative intervention, a creative media project such as the one we are outlining here has the potential to become a new incarnation of the age-old "theater-within-the theater" device, whose actors are at the same time critics.

For Judith Butler, when drawing on Foucault's work, the critic has a double task: to show how knowledge and power work to constitute a more or less systematic way of ordering the world with its own "conditions of acceptability of a system" but also "to follow the breaking points which indicate its emergence." So not only is it necessary to isolate and identify the peculiar nexus of power and knowledge that gives rise to the field of intelligible things, but also to track the way in which that field meets its breaking point, the moments of its discontinuities, the sites where it fails to constitute the intelligibility for which it stands.[4] It was with this is mind that we showed in chapter 4 how an industry vision of Ambient Intelligence, entangled with noncelebratory science fiction, manages to both make and unmake itself, to write itself and write itself off, while constituting an open invitation to other writers, artists, technologists, and designers to intervene and join in. "Creative media" can therefore perhaps be seen as one of the emergent paradigms at the interfaces of performativity and vitality that this book is aiming to map out. What can hopefully emerge through this playful yet rigorous cross-disciplinary approach will be a more dynamic, networked, and engaged mode of working on and with "the media," in which critique is always already accompanied by the work of participation and invention.[5]

With this creative media project, we are thus hoping to stage a new paradigm not only for *doing media critique as media analysis* but also for *inventing (new) media*. The reason for this is that, as we have argued in chapter 1, conventional forms of media analysis are ineffective in as far as they are based on what we perceive to be a set of false problems and false divisions. The false problems involve current conceptions of interactivity, convergence, determinism, constructionism, information, and identity. False divisions, which continue to structure debates on new media in particular, include those between production and consumption, text and image, and language and materiality. We have also maintained that there is no rigid division between new and old media, as ongoing processes of differentiation are constantly taking place across *all media*. The underlying problem of "the media" is precisely that of *mediation*: of the economic, cultural, social, technical, textual, political, and psychological processes through which a variety of media forms continue to develop and stabilize over time.

The problem of mediation is for us both contingent and temporal. It centers on the evolution of media in relation to wider socioeconomic forces. The role of technology in this process of evolution is neither determining nor determined. Neither is this role merely instrumental or anthropological; rather, it is *vital* and *relational*. If the essence of technology is inseparable from the essence of humanity, as Heidegger has it, then there is no justification for positing humanism against technicism, or vice versa. There is also no point in fighting "against technology." But there is every point—or, indeed, even an ethicopolitical injunction—in exploring practices of differentiation at work in the current mediascape: globalized "media events," TV makeover shows, intelligent media devices. *Our creative media project thus seeks to promote the invention of different forms of engagement with media.* This is not to say that differentiation is always welcome and beneficial, and that all forms of difference are to be equally desired, no matter what material and symbolic effects they generate. Our emphasis is therefore on creative-critical practices that are neither simply oppositional nor consensual, and that attempt, in Donna Haraway's words, to "make a difference" within processes of mediation. To put it another way, we are interested in staging interventions across conventional boundaries of theory and practice, art and industry, the humanities and the sciences. Such interventions may come to constitute events that cannot be determined a priori.

With a nod to Samuel Beckett,[6] the key points of our manifesto can thus be summarized as follows:

1. All of old. But ask again. Ask better.

We are seeking to unsettle the comforting certainties that are assumed by sociologists or historians of "new" media landscapes who purport to describe but not create them, who survey the mediascape and its objects without seeing themselves as a constitutive part of them. Following Karen Barad (but not all the way), we aim to expose *as illusions* the ontological gaps that allow media events to have social effects that might be good or bad, celebratory or catastrophic. Taking Barad back to Bergson, we want to stress not only our lively relationalities of becoming-with-media but also the importance of asking better questions about them—as opposed to seeking ready-made, predetermined answers that might come, say, from a technicist or a humanist perspective. So we ask, for example, what is *really* at stake in the Large Hadron Collider project? Is it the discovery and representation of the elusive (invisible) God particle that has not appeared yet and that we are still eagerly awaiting? Or is it rather something that Baudrillard offers us insight into—namely, the staging of the relation between catastrophic or celebratory events and their mediation?[7]

2. Cut. Cut again. Cut better.

We seek to interrupt the equally seductive logic of media and technological flows and say, following Derrida: Stop! Cut! There are *not* only processes. We should not—contrary to what E. M. Forster suggests—"only connect."[8] The ethical possibilities of media such as, say, photography are predicated on differentiations that *are made* and that do not *just happen*. Derrida's transcendent cut cuts across Bergson's immanence with a stronger imperative for us to account for, and show hospitality toward, our technological artifacts and media objects. That does not mean it is all about us: the things that we are making are also making us. Agential cuts, as Barad argues, cut both ways. This is not a paranoid observation, a twenty-first-century take on the fear of the machine, that the tools we use may well be using us. Such paranoid responses are predicated on humanism's autonomies—and on the anthropological and instrumental approach to technology that tells us little about what technology is. But paranoia will not get us very far. It is precisely in the face of such reductive instrumentalism—the enframing of media, technologies, and subjects within a totalizing neoliberal rationality—that we find ourselves, alongside our feminist and other philosophical companions, tilting at windmills and determined to make a difference.

3. Try again. Fail again. Fail better.

In the spirit of critical and creative experimentation, and in recognition of the inevitable indeterminacy of our interventions into the possibilities of life after new media, we borrow from Beckett—a writer who played with a Baudrillard type of endgame, and who changed the game in the process—an injunction *to take a risk in the face of the unknown and the impossible, the not-even-yet-new, against all odds.*

Critics of the world, create![9]

Notes

Introduction

1. Although we will argue throughout this volume, and in chapter 2 in particular, for the configuration of media and liveness, mediation, and life, we will also seek to avoid Bergson's "false division" between life and matter. This division is "false," or problematic, because it enables him to abstract life from material and symbolic forms—including, in our case, media forms. Our claim is that although media continue to stake their claim to liveness (live TV, live Twitter feeds, etc.), by virtue of being inseparable (though different *in kind*) from processes of mediation, media are co-constitutive of life itself—which, under certain circumstances, and through a sequence of reductionist operations, can subsequently also take a media form (a CD with one's genetic profile; synthetic biology database, etc.).

2. See Benjamin Peters, "And Lead Us Not into Thinking the New is New: A Bibliographic Case for New Media History," *New Media and Society*, 11, no. 1–2 (2009): 13–30.

3. CCD stands for "charge-coupled device," in which electrical charge can be manipulated to obtain translation of signal into digital value. It is frequently used in digital cameras.

4. Lev Manovich, *The Language of New Media* (Cambridge, MA: MIT Press, 2001), 25.

5. Wendy Hui Kyong Chun, "Introduction: Did Somebody Say New Media?" In *New Media, Old Media*, ed. Wendy Hui Kyong Chun and Thomas Keenan (New York: Routledge, 2006), 9.

6. Chun, "Introduction," 3.

7. See, for instance, Sarah Kember, *Cyberfeminism and Artificial Life* (London: Routledge, 2003), and Joanna Zylinska, *Bioethics in the Age of New* Media (Cambridge, MA: MIT Press, 2009).

8. The term "lifeness" we propose in the book is aimed to go beyond what we will argue is quite a static view of media as espoused by terms such as "live TV," "live news," and so on, in an attempt to convey what we see as the dynamic vitality of mediation processes.

9. Rosi Braidotti, *Nomadic Subjects: Embodiment and Sexual Difference in Contemporary Feminist Theory* (New York: Columbia University Press), 17.

10. Bernard Stiegler, *Technics and Time, 2: Disorientation*, trans. Stephen Barker (Stanford: Stanford University Press, 2009), 186.

11. Lucy Suchman, *Human–Machine Reconfigurations* (Cambridge: Cambridge University Press, 2007), 206.

12. Rosi Braidotti, *Transpositions* (Cambridge: Polity Press, 2006), 6.

Chapter 1

1. See Giles Deleuze, *Bergsonism* (New York: Zone Books, 1998), 13–35.

2. See Gary Hall, *Digitize This Book! The Politics of New Media, or Why We Need Open Access Now* (Minneapolis: University of Minnesota Press, 2008), 210, 214.

3. Joost Van Loon's book *Media Technology: Critical Perspectives* (Maidenhead, UK: Open University Press, 2007) represents for us precisely such an example of an unfulfilled promise to engage in a serious and rigorous manner with the notion of mediation.

4. The comparison here is between media and mediation. If we apply the same method to a single medium such as photography (see Sarah Kember, "The Virtual Life of Photography," *Photographies*, 1, no. 2 (September 2008): 175–203), we can investigate its specific ontology or identify it as a difference in kind from other media.

5. In her critique of dominant narratives about the media, Gitelman points out that media "tend unthinkingly to be regarded as heading a certain 'coherent and directional' way along an inevitable path, a history, toward a specific and not-so-distant end. Today, the imagination of that end point in the United States remains uncritically replete with confidence in liberal democracy and has been most uniquely characterized by the cheerful expectation that digital media are all converging toward some harmonious combination or global 'synergy,' if not also toward some perfect reconciliation of 'man' and machine" in *Always Already New: Media, History, and the Data of Culture* (Cambridge, MA: MIT Press), 3.

6. Hall, *Digitize This Book!*, 161.

7. Martin Lister makes a similar argument about photography in his introduction to his edited volume *The Photographic Image in Digital Culture* (London: Routledge, 2005). He argues that photography, always a heterogeneous medium with a complex and problematic relation to the truth, was assigned a more homogeneous and significantly less complex status with the advent of digitization and the presumed effects the latter was having on photographic truth.

8. David Teather, "End of Story for Hardbacks? Amazon's Ebook Milestone," *The Guardian*, July 21, 2010.

9. Stanislaw Lem, *Summa Technologiae* (Lublin: Wydawnictwo Lubelskie, 1984), 8. An English translation of this book by Joanna Zylinska is forthcoming from the University of Minnesota Press in 2012.

10. Alan Turing, "Computing Machinery and Intelligence," *Mind* 59 (October 1950): 433–460.

11. Fred Ritchin, "The End of Photography As We Have Known It," in *Photovideo: Photography in the Age of the Computer*, ed. Paul Wombell (London: Rivers Oram Press, 1991), 8–15. For a discussion of photography and digitization, see Sarah Kember, *Virtual Anxiety. Photography, New Technology and Subjectivity* (Manchester: Manchester University Press, 1998).

12. Jay David Bolter and Richard Grusin, *Remediation: Understanding New Media*, (Cambridge, MA: MIT Press, 2000), 187.

13. Nicholas Carr, *The Shallows: How the Internet Is Changing the Way We Think, Read and Remember* (New York: WW Norton & Co, 2010).

14. Martin Lister, Jon Dovey, Seth Giddings, Iain Grant, and Kieran Kelly, *New Media: A Critical Introduction*, 2nd ed. (New York: Routledge, 2008), 4.

15. Lister et al., *New Media*, 77.

16. See Bolter and Grusin, *Remediation*, 19; Lister et al., *New Media*, 95. This point is particularly important when we consider the degree to which media are recombined not only with information but with biotechnologies (see Sarah Kember, "Doing Technoscience as (New) Media," in *Media and Cultural Theory*, ed. James Curran and David Morley [New York: Routledge, 2006]).

17. Lister et al., *New Media*, 94.

18. Lister et al., *New Media*, 95.

19. Bolter and Grusin, *Remediation*, 9.

20. Lem, *Summa Technologiae*, 205.

21. Bolter and Grusin, *Remediation*, 5.

22. Bolter and Grusin, *Remediation*, 25–26.

23. Bolter and Grusin, *Remediation*, 26.

24. Bolter and Grusin, *Remediation*, 14.

25. Bolter and Grusin, *Remediation*, 14.

26. Bolter and Grusin, *Remediation*, 55.

27. Bolter and Grusin, *Remediation*, 55, emphasis in original.

28. Bolter and Grusin, *Remediation*, 59.

29. Bolter and Grusin, *Remediation*, 58n7.

30. Carolyn Marvin, *When Old Technologies Were New* (New York: Oxford University Press, 1988), 3.

31. Marvin, *When Old Technologies*, 8.

32. Marvin, *When Old Technologies*, 8.

33. Gitelman, *Always Already New*, xi.

34. Gitelman, *Always Already New*, 9.

35. See Gitelman, *Always Already New*, 8.

36. Katherine N. Hayles, *My Mother Was a Computer: Digital Subjects and Literary Texts* (Chicago: University of Chicago Press, 2005), 33.

37. See Karen Barad, *Meeting the Universe Halfway* (Durham, NC: Duke University Press, 2007).

38. See, for example, Matthew Fuller, *Media Ecologies: Materialist Energies in Art and Technoculture* (Cambridge, MA: MIT Press, 2005).

39. Examples of media and communications books that expound such a spatial paradigm include Manuel Castells, *The Rise of the Network Society* (Oxford: Blackwell, 1996), and Nick Couldry and Anna McCarthy (eds.), *MediaSpace: Place, Scale and Culture in a Media Age* (New York: Routledge, 2003).

40. "Today, the instantaneous world of electric information media involves all of us, all at once. Ours is a brand new world of. Time, in a sense, has ceased and space has vanished," McLuhan on McLuhanism, WNDT Educational Broadcasting Network, 1966.

41. Lister et al., *New Media*, 95.

42. Lister et al., *New Media*, 94.

43. Some of the ideas on Bernard Stiegler's work included in this section of the chapter were originally developed in Joanna Zylinska, "Playing God, Playing Adam: The Politics and Ethics of Enhancement," *Journal of Bioethical Inquiry* 7, no. 2 (2010): 149–161.

44. Mark Poster, "High-Tech Frankenstein, or Heidegger Meets Stelarc," in *The Cyborg Experiments: The Extensions of the Body in the Media Age*, ed. Joanna Zylinska (London: Continuum, 2002), 20.

45. Martin Heidegger, "The Question Concerning Technology," in Martin Heidegger, *Basic Writings*, ed. David Farrell Krell (New York: Harper & Row, 1977), 304.

46. Heidegger, "The Question Concerning Technology," 293.

47. Heidegger, "The Question Concerning Technology," 293.

48. Poster, "High-Tech Frankenstein," 20.

49. Poster, "High-Tech Frankenstein," 20.

50. Heidegger, "The Question Concerning Technology," 287.

51. Heidegger, "The Question Concerning Technology," 287.

52. Heidegger, "The Question Concerning Technology," 295.

53. Poster, "High-Tech Frankenstein," 18.

54. Heidegger, "The Question Concerning Technology," 288.

55. Heidegger, "The Question Concerning Technology," 299.

56. Heidegger, "The Question Concerning Technology," 300.

57. Heidegger, "The Question Concerning Technology," 300.

58. For an exposition of this notion of ethics as a response to the alterity of the other, see Emmanuel Levinas, "The Trace of the Other," in *Deconstruction in Context*, ed. Mark C. Taylor (Chicago: University of Chicago Press, 1986), and Joanna Zylinska, "A User's Guide to Culture, Ethics and Politics," in *The Ethics of Cultural Studies* (London: Continuum, 2005).

59. Bernard Stiegler, *Technics and Time, 1: The Fault of Epimetheus*, trans. Richard Beardsworth and George Collins (Stanford: Stanford University Press, 1998), 187, emphasis in original.

60. Stiegler, *Technics and Time, 1*, 187.

61. Cited in Stiegler, *Technics and Time, 1*, 188.

62. *The Ister* (2004), directed by David Barison and Daniel Ross.

63. It has to be acknowledged that Stiegler's work runs against some of the very same "humanist" limitations that his work is frequently used to critique. This is evident in the way he reintroduces a number of problematic anthropological distinctions such as those between culture and nature, or human and animal, into his argument outlined in *Technics and Time*.

64. Heidegger, "The Question Concerning Technology," 309.

65. Ethics, for Levinas, is not something imposed from outside or above; instead, ethics is inevitable. An ethical event occurs in every encounter with difference, with the "face" and discourse of the other that addresses me and makes me both responsible and accountable (even if I ultimately decide to turn my back on this difference or even annihilate it). I am thus always already a hostage of the other, of his or her ethical demand.

66. Stiegler, *Technics and Time, 1*, 128.

67. This point has been argued before by Joanna Zylinska in *Bioethics in the Age of New Media* (Cambridge, MA: MIT Press, 2009, x–xi).

68. Lev Manovich, *The Language of New Media* (Cambridge, MA: MIT Press, 2001), 19.

69. Lister et al., *New Media*, 2.

70. Hall, *Digitize This Book!*, 214.

71. http://en.wikipedia.org/wiki/Mediation_%28Marxist_theory_and_media_studies%29.

72. Aeron Davis, *The Mediation of Power: A Critical Introduction* (London: Routledge, 2007), 13.

73. Davis, *The Mediation of Power*, 14.

74. Nick Couldry, "Mediatization or Mediation? Alternative Understandings of the Emergent Space of Digital Storytelling," *New Media and Society* 10, no. 3 (2008): 373–391, 379.

75. Couldry, "Mediatization or Mediation?" 379.

76. The concept of mediation has made an appearance in various publications on new media (Leah A. Lievrouw and Sonia Livingstone [eds.], *The Handbook of New Media* (London: Sage, 2006); Van Loon, *Media Technology*), but it has so far remained rather underdeveloped, with the concerns of the mediation "process" just being seen as a function of new media "objects."

77. Cited in Couldry, "Mediatization or Mediation?" 380.

78. Couldry, "Mediatization or Mediation?" 381.

79. Couldry describes mediation as "capturing a variety of dynamics within media flows" in "Mediatization or Mediation?" 380.

80. Heidegger, "The Question Concerning Technology," 293.

81. Eugene Thacker, "What Is Biomedia?," *Configurations* 11.1 (2003): 47–79.

82. Scott Lash and Celia Lury, *Global Culture Industry: The Mediation of Things* (Cambridge: Polity, 2007), 111.

83. Gary Gumpert and Robert Cathcart, "A Theory of Mediation," in *Mediation, Information and Communication*, ed. B. Ruben and L. Lievrouw (New Brunswick, NJ: Transaction, 1990), 23.

84. Gumpert and Cathcart, "A Theory of Mediation," 23.

85. Gumpert and Cathcart, "A Theory of Mediation," 24.

86. Gumpert and Cathcart, "A Theory of Mediation," 26.

87. Gumpert and Cathcart, "A Theory of Mediation," 26.

88. Gumpert and Cathcart, "A Theory of Mediation," 35.

89. Gumpert and Cathcart, "A Theory of Mediation," 31.

90. Yet the limitation of Gumpert and Cathcart's theory is ultimately revealed in their return to (and of) the human through the humanist backdoor in their media ecology, as is evident in their declaration that media transformation comes from the creative impulse in humans (or, as they put it, "the impulse to extend and replicate personal experiences" in "A Theory of Mediation," 27), without taking into account the creative force of media themselves. Now, the recognition of the human singularity within a given media ecology is not a problem per se, both in terms of the affordances and effectivities it opens up, and the philosophical reflection it allows for, provided that it is not confused with human primacy or species supremacy. Yet this singularity of the human and of human agency has to be reconciled with the multiple singularity of media and with the interwoven processes of mediation.

91. Sarah Kember, "The Virtual Life of Photography," *Photographies* 1, no. 2 (September 2008): 180. For a discussion of the foreclosure of creativity in evolutionary art and games, see also Sarah Kember, "Metamorphoses: The Myth of Evolutionary Possibility," in *Inventive Life: Approaches to the New Vitalism*, ed. Mariam Fraser, Sarah Kember, and Celia Lury (London Sage Publications, 2005).

92. Deleuze, *Bergsonism*, 32. Key elements of our approach to Bergson were outlined in Sarah Kember, "Creative Evolution? The Quest for Life (on Mars)," *Culture Machine: InterZone* (March 2006), http://www.culturemachine.net/index.php/cm/article/view/235/216. Bergson's intuitive method is applied to the question of the nonhuman and to the question of feminist ethics in

Sarah Kember, "No Humans Allowed? The Alien in/as Feminist Theory," *Feminist Theory* 12, no. 2 (2011): 183–199.

93. As Michael Hardt puts it, "Affects require us, as the term suggests, to enter the realm of causality, but they offer a complex view of causality because the affects belong simultaneously to both sides of the causal relationship. They illuminate, in other words, both our power to affect the world around us and our power to be affected by it," in "Foreword: What Affects are Good For," in *The Affective Turn: Theorizing the Social*, ed. Patricia Ticineto Clough with Jean Halley (Durham, NC: Duke University Press, 2007), ix.

94. Mechanism for Bergson is an approach within science and evolutionary theory in which nature is regarded as a machine "regulated by mathematical laws" (Bergson, *Creative Evolution*, trans. Arthur Mitchell [New York: Random House, The Modern Library, 1941], 51). Finalism tends to see reality as the realization of a plan. For Bergson, these approaches converge in that they are both "reluctant to see in the course of things generally, or even simply in the development of life, an unforeseeable creation of form" (51).

95. Deleuze, *Bergsonism*, 13.

96. See Elizabeth Grosz, *The Nick of Time: Politics, Evolution and the Untimely* (Durham: Duke University Press, 2004), 237.

97. This phrase is derived from Christopher G. Langton's paper "Artificial Life," in *The Philosophy of Artificial Life*, ed. Margaret Boden (Oxford and New York: Oxford University Press, 2006). The possibilities of artificial life are examined in Sarah Kember, *Cyberfeminism and Artificial Life* (New York: Routledge, 2003).

98. Henri Bergson, *The Creative Mind: An Introduction to Metaphysics* (New York: Kensington Publishing, 2002), 31.

99. See Bergson, *The Creative Mind*, 51.

100. See Bergson, *The Creative Mind*, 33.

101. Jacques Derrida, "Artifactualities," in Jacques Derrida and Bernard Stiegler, *Echographies of Television*, trans. Jennifer Bajorek (Cambridge: Polity Press, 2002), 11.

102. Bergson, *The Creative Mind*, 50.

103. Bergson, *The Creative Mind*, 83.

104. Bergson, *The Creative Mind*, 32.

105. Jacques Derrida, "Différance," *Margins of Philosophy*, trans. Alan Bass (Chicago: University of Chicago Press, 1982), 11.

106. See Hall, *Digitize This Book!*, 208–215, and Joanna Zylinska, *Bioethics in the Age of New* Media (Cambridge, MA: MIT Press, 2009), 30.

107. Derrida, "Artifactualities," 7.

108. See Derrida, "Différance."

109. See Daniel W. Smith, "Deleuze and Derrida, Immanence and Transcendence: Two Directions in Recent French Thought," in *Between Deleuze and Derrida*, ed. Paul Patton and John Protevi (London: Continuum, 2003), 61.

110. Paul Patton and John Protevi, "Introduction," in *Between Deleuze and Derrida*, ed. Paul Patton and John Protevi (London and New York: Continuum, 2003), 22.

111. Derrida, "Différance," 3.

112. Derrida, "Différance," 7.

113. Derrida, "Différance," 7.

Chapter 2

1. Cited in Larry Elliott, "Analysis: Crunch Time for Germany," *The Guardian*, November 16, 2011, 15.

2. Carl Wilkinson, *The Observer Book of Space* (London: Observer Books, 2007), 112.

3. See Karen Barad, *Meeting the Universe Halfway* (Durham, NC: Duke University Press, 2007), 49.

4. William Merrin, *Baudrillard and the Media* (Cambridge: Polity Press, 2005), 74.

5. Merrin, *Baudrillard and the Media*, 74.

6. See Jay David Bolter and Richard Grusin, *Remediation: Understanding New Media*, Cambridge, MA: MIT Press, 2000); Joost Van Loon, *Media Technology: Critical Perspectives* (Maidenhead, England: Open University Press, 1997).

7. See Joost Van Loon, "Modalities of Mediation," in *Media Events in a Global Age*, ed. Nick Couldry, Andreas Hepp, and Friedrich Krotz (London: Routledge, 2010): 109–123.

8. Barad, *Meeting the Universe*, 46.

9. See Henri Bergson, *Creative Evolution* (New York: Random House, 1941) and *The Creative Mind* (New York: Kensington Publishing, 1992).

10. Jean Baudrillard, *Simulations* (New York: Semiotext(e), 1983), 3.

11. See Baudrillard, *Simulations*.

12. We are recognizing this question as an essentially Baudrillardian one, yet we are seeking to reexamine it here, not least through comparing and contrasting the terms "simulation" and "mediation."

13. Daniel Dayan and Elihu Katz, *Media Events: The Live Broadcasting of History* (Cambridge, MA: Harvard University Press, 1992), 32.

14. Nick Couldry, *Media Rituals: A Critical Approach* (London: Routledge, 2003), 61.

15. See Edward A. Shils and Michael Young (1975 [1956]) "The Meaning of the Coronation," in *Center and Periphery: Essays in Macrosociology*, ed. Edward Shils (Chicago: University of Chicago Press, 1975).

16. See Daniel J. Boorstin, *The Image: A Guide to Pseudo-Events in America* (New York: Harper Colophon, 1961).

17. Dayan and Katz, *Media Events*, viii.

18. See Émile Durkheim, *The Elementary Forms of the Religious Life* (New York: The Free Press, 1968).

19. Dayan and Katz, *Media Events*, 1.

20. Dayan and Katz, *Media Events*, 8.

21. Dayan and Katz, *Media Events*, 8.

22. Daniel Dayan, "Beyond Media Events: Disenchantment, Derailment, Disruption," in *Media Events in a Global Age*, ed. Nick Couldry, Andreas Hepp and Friedrich Krotz (London: Routledge, 2010), 25.

23. Dayan, "Beyond Media Events," 25.

24. Dayan, "Beyond Media Events," 25.

25. See Andreas Hepp and Nick Couldry, "Introduction: Media Events in Globalized Media Cultures," in Couldry, Hepp, and Krotz, *Media Events*.

26. Dayan, "Beyond Media Events," 25.

27. Dayan, "Beyond Media Events," 25.

28. Elihu Katz and Tamar Liebes,'"NO MORE PEACE!': How Disaster, Terror and War Have Upstaged Media Events," in Couldry, Hepp, and Krotz, *Media Events,* 33.

29. Katz and Liebes,'"NO MORE PEACE!,'" 33.

30. Katz and Liebes,'"NO MORE PEACE!,'" 33.

31. See Andreas Hepp and Nick Couldry, "Introduction: Media Events in Globalized Media Cultures," in Couldry, Hepp, and Krotz, *Media Events*.

32. See Hepp and Couldry, "Introduction."

33. The documentary, for example, makes continual reference to the particulate matter of which we are composed. It refers to the relation between particles, mass, matter, and bodies that is theorized in the field of particle physics. See Frank Wilczek, *The Lightness of Being: Mass, Ether, and the Unification of Forces* (New York: Basic Books, 2008).

34. In his introduction to *Forget Foucault,* Sylvère Lotringer writes: "Baudrillard didn't ground his theory anywhere, except *on the very process of deterritorialization*. On the principle of death itself" ("Introduction: Exterminating Angel," in *Forget Foucault,* ed. Jean Baudrillard [Los Angeles: Semiotext(e), 2007], 17).

35. Dayan, "Beyond Media Events," 29.

36. Dayan, "Beyond Media Events," 25.

37. Dayan and Katz, *Media Events*, 23.

38. Dayan, "Beyond Media Events," 25.

39. Dayan, "Beyond Media Events," 26.

40. See Katz and Liebes,'"NO MORE PEACE!.'"

41. Dayan, "Beyond Media Events," 40. For Baudrillard, simulation is "opposed to representation." Simulation "envelops the whole edifice of representation as itself a simulacrum" (*Simulations*, 11). We will argue that mediation, in contrast to simulation, does not envelop representation but rather incorporates and exceed it.

42. Katz and Liebes, '"NO MORE PEACE!,'" 33.

43. Katz and Liebes, '"NO MORE PEACE!,'" 34.

44. Katz and Liebes, '"NO MORE PEACE!,'" 35.

45. Katz and Liebes, '"NO MORE PEACE!,'" 35.

46. Katz and Liebes, '"NO MORE PEACE!,'" 35.

47. Katz and Liebes, '"NO MORE PEACE!,'" 39.

48. Hepp and Couldry, "Introduction," 11.

49. Couldry, *Media Rituals*, 4.

50. Couldry, *Media Rituals*, 4.

51. Couldry, *Media Rituals*, 56.

52. Couldry, *Media Rituals*, 66.

53. In *The Creative Mind* (New York: Kensington Publishing, 1992), Bergson cautions us against the division between realism and idealism on the basis that they are effectively two sides of the same coin.

54. See J. L. Austin, *How To Do Things with Words* (Cambridge, MA: Harvard University Press, 1962); Jacques Derrida, *Limited Inc.* (Evanston: Northwestern University Press, 1988); Judith Butler, *Excitable Speech: A Politics of the Performative* (New York: Routledge, 1997); Karen Barad, *Meeting the Universe Halfway* (Durham, NC: Duke University Press, 2007).

55. Van Loon, "Modalities of Mediation," 114.
56. Van Loon, "Modalities of Mediation," 114.
57. Couldry, *Media Rituals*, 1, 18.
58. Van Loon, "Modalities of Mediation," 114.
59. Merrin, *Baudrillard and the Media*, 64.
60. Jean Baudrillard, *The Gulf War Did Not Take Place* (Sydney: Power Publications, 2009).
61. Merrin, *Baudrillard and the Media*, 72.
62. Baudrillard, *Simulations*, 4.
63. Baudrillard, *Simulations*, 65.
64. Baudrillard, *Forget Foucault*, 98.
65. Baudrillard, *Forget Foucault*, 98.
66. Baudrillard, *Simulations*, 57.
67. Merrin, *Baudrillard and the Media*, 66, emphasis added.
68. Merrin, *Baudrillard and the Media*, 76, emphasis in original.
69. Merrin, *Baudrillard and the Media*, 77.
70. Jean Baudrillard, *The Transparency of Evil: Essays on Extreme Phenomena* (London: Verso, 2002), 26, emphasis added.
71. Baudrillard, *The Transparency of Evil*, 26.
72. Baudrillard, *The Transparency of Evil*, 27.
73. Baudrillard, *The Transparency of Evil*, 27.
74. Baudrillard, *The Transparency of Evil*, 28.
75. Baudrillard, *Simulations*, 26.
76. Van Loon, "Modalities of Mediation," 115.
77. Van Loon, "Modalities of Mediation," 111–112.
78. Van Loon, "Modalities of Mediation," 117.
79. See Rosi Braidotti, *Transpositions* (Cambridge: Polity Press, 2006).
80. Van Loon, "Modalities of Mediation," 111.
81. Katz and Liebes,'"NO MORE PEACE!,'" 35.
82. Michael Seamark, "Does This BBC Man Have Too Much Power? Reporter Blamed for Helping Trigger Shares Fall," *Mail Online* (October 8, 2008), accessed July 7, 2011, http://www.dailymail.co.uk/news/article-1072549/BBC-reporter-Robert-Peston-blamed-helping-trigger-shares-fall.html.
83. See Alex Brummer, *The Crunch: The Scandal of Northern Rock and the Escalating Credit Crisis* (London: Random House Business Books, 2008); Graham Turner, *The Credit Crunch* (London: Pluto, 2008).
84. See Robert Peston, "The New Capitalism," BBC NEWS (December 8, 2008), accessed July 7, 2011, http://www.bbc.co.uk/blogs/thereporters/robertpeston/newcapitalism.pdf.
85. Goldman Sachs denied an accusation of fraud made by the US regulators. It is said that they made "misleading statements and omissions in connection with a synthetic collateralized debt obligation" (Richard Adams, "Selling Fire Insurance to Arsonists," *Guardian.co.uk* (April 20, 2010), accessed July 7, 2011, http://www.guardian.co.uk/commentisfree/cifamerica/2010/apr/20/goldman-sachs-sec-fraud).
86. Brummer, *The Crunch*, 27.

87. Dayan and Katz, *Media Events*, 23.

88. See Jane Feuer, "The Concept of Live Television: Ontology as Ideology," in *Regarding Television*, E. Ann Kaplan, ed. (Frederick: University Publications of America, 1983).

89. Jacques Derrida, "Artifactualities," in *Echographies of Television*, ed. Jacques Derrida and Bernard Stiegler, trans. Jennifer Bajorek (Cambridge: Polity Press, 2002), 40.

90. Derrida, "Artifactualities," 40.

91. Derrida, "Artifactualities," 3.

92. Derrida, "Artifactualities," 3.

93. Derrida, "Artifactualities," 3.

94. Derrida, "Artifactualities," 5.

95. Derrida, "Artifactualities," 10. He writes further on that page: "The event, the singularity of the event, that's what différance is all about."

96. Derrida, "Artifactualities," 11.

97. Derrida, "Artifactualities," 6.

98. Henri Bergson, *The Creative Mind*, 15.

99. As Laurie Anderson sings in "Sharkey's Day": "Hey! Look out! Bugs are crawling up my legs! You know? I'd rather see this on TV. Tones it down."

100. Feuer, "The Concept of Live Television," 13.

101. Mary Ann Doane, "Information, Crisis, Catastrophe," in *Logics of Television*, Patricia Mellencamp, ed. (Bloomington: Indiana University Press, 1990), 22.

102. Doane, "Information, Crisis, Catastrophe," 228–231.

103. See, for example, Doane, "Information, Crisis, Catastrophe"; Constance Penley, *NASA/Trek: Popular Science and Sex in America* (New York and London: Verso, 1997); Mimi White, "The Attractions of Television: Reconsidering Liveness," in *Media Space: Place, Scale and Culture in a Media Age*, Nick Couldry and Anna McCarthy, eds. (London and New York: Routledge, 2003).

104. See Katz and Liebes,'"NO MORE PEACE!"'; Baudrillard, *The Transparency of Evil*.

105. See White, "The Attractions of Television."

106. White, "The Attractions of Television," 76.

107. Cited in White, "The Attractions of Television," 76.

108. Penley, *NASA/Trek*, 77.

109. Penley, *NASA/Trek*, 77.

110. Penley, *NASA/Trek*, 77.

111. Penley, *NASA/Trek*, 77.

112. Penley, *NASA/Trek*, 79.

113. Penley, *NASA/Trek*, 87.

114. Penley, *NASA/Trek*, 87.

115. See Andrew Pierce, "The Queen Asks Why No One Saw the Credit Crunch Coming," *The Telegraph* (November 5, 2008), accessed July 7, 2011, http://www.telegraph.co.uk/news/uknews/theroyalfamily/3386353/The-Queen-asks-why-no-one-saw-the-credit-crunch-coming.html.

116. Brummer, *The Crunch*, 17.

117. Robert Peston, speaking on BBC NEWS on September 13, 2007.

118. BBC News, "Northern Rock Gets Bank Bail Out" (September 13, 2007), accessed July 7, 2011, http://news.bbc.co.uk/1/hi/business/6994099.stm.

119. Brummer, *The Crunch*, 80.

120. Brummer, *The Crunch*, 81.

121. Securitization refers to the "creation of asset-backed securities supported by a stream of cash flow" (Brummer, *The Crunch*, 81). In other words, it is the practice through which "banks package up mortgage loans and sell them on" (16).

122. Brummer, *The Crunch*, 81–82, emphasis added.

123. But for the curious, securitization is defined earlier. Debt bundling refers to much the same phenomenon: this is where mortgage loans are bundled up and sold on to investors. A junk bond is "a high-risk, low-credit-rated, non-investment grade bond" (Brummer, *The Crunch*, 232). Liquidity is "the ability to convert an asset into cash through buying or selling" (232). Hedge funds are privately owned investment companies. A vanilla loan is a basic or standard financial package, and quantitative easing is a way of increasing the flow of money within an economy.

124. The merger of HBOS and Lloyds TSB took place in September 2008.

125. Peston announced that three of the United Kingdom's largest banks—Barclays, Royal Bank of Scotland, and Lloyds TSB—had gone to the Treasury for financial support.

126. Peston is referring to a report he made in 1997 "that the newly elected Labour government was contemplating joining the euro within that parliament" (Seamark, "Does This BBC Man Have Too Much Power?").

127. Mark Fenton O'Creavy, "Has Robert Peston Caused a Recession? Social Amplification, Performativity and Risks in Financial Markets" (October 17, 2008), accessed July 7, 2011, http://www.open2.net/blogs/money/index.php; emphasis in original.

128. Donald Mackenzie, Fabian Muniesa, and Lucia Siu Callon, "Introduction," in *Do Economists Make Markets?*, Donald Mackenzie, Fabian Muniesa, and Lucia Siu, eds. (Princeton and Oxford: Princeton University Press, 2007), 2.

129. Mackenzie et al., "Introduction," 2.

130. Ian Hacking, *Representing and Intervening: Introductory Topics in the Philosophy of Natural Science* (Cambridge: Cambridge University Press, 1983), 3.

131. Hacking, *Representing and Intervening*, 5.

132. Hacking, *Representing and Intervening*, 321.

133. Hacking, *Representing and Intervening*, 315.

134. Michel Callon, "What Does It Mean to Say That Economics Is Performative?" in Mackenzie, Muniesa, and Siu, *Do Economists Make Markets?*, 329.

135. Callon, "What Does It Mean to Say That Economics Is Performative?," 335.

136. Callon, "What Does It Mean to Say That Economics Is Performative?," 335.

137. Callon, "What Does It Mean to Say That Economics Is Performative?," 329.

138. Brummer, *The Crunch*, 81.

139. Callon, "What Does It Mean to Say That Economics Is Performative?," 323.

140. See Barad, *Meeting the Universe Halfway*; Mark B. N. Hansen, *New Philosophy for New Media* (Cambridge, MA: MIT Press, 2004).

141. See Callon, "What Does It Mean to Say That Economics Is Performative?"; Andrew Pickering, *The Mangle of Practice: Time, Agency & Science* (Chicago: The University of Chicago Press, 1995).

142. Dayan, "Beyond Media Events, 25.

143. Barad, *Meeting the Universe*, 46.

144. Barad, *Meeting the Universe*, 46.

145. Barad, *Meeting the Universe*, 46.

146. As such, it also offers an indirect comment on Baudrillard's recourse to the semiotics of signs in this theory of simulation. The science of signs is applicable only to the field of representations, which, for Baudrillard, has disappeared, suggesting that semiotics is therefore rendered redundant. Moreover, semiotics is (post)structural and atemporal and therefore not particularly well suited to the elucidation of material agential processes such as mediation. For Pickering, the "alternative to the invocation of an atemporal science of signs is to think carefully about time" (*The Mangle of Practice*, 14). This statement has led us to put Baudrillard's thinking into conversation with that of Bergson as well as Barad in this chapter.

147. http://lhc-machine-outreach.web.cern.ch.

148. "Event" is the term used in particle physics to describe a collision. On Tuesday, March 30, 2010, collisions occurred at half of the energy that the LHC project originally aimed for, namely 7 TeV (teraelectronvolts). Although hailed as a success, this is still part of an experimental period, after which the LHC will be shut down prior to making its final, full-energy attempt.

149. Barad, *Meeting the Universe Halfway*, 9.

150. Only broadly, because of the sometimes unresolved distinction here between life and matter. For a discussion of this point, see Sarah Kember, "Creative Evolution? The Quest for Life (on Mars)," *Culture Machine: Interzone* (March 2006), accessed July 7, 2011, http://www.culturemachine.net/index.php/cm/article/view/235/216.

151. This outline of Bergson's "method" is provided by Deleuze in his book *Bergsonism* (New York: Zone Books, 1988).

152. John Ellis outlines some of the "fundamental questions" that remain unanswered in physics. These include: "the origin of particle masses, the small difference between matter and antimatter . . . the origin of matter, and the natures of dark matter and dark energy" ("The Fundamental Physics Behind the LHC," in *The Large Hadron Collider: A Marvel of Technology*, ed. Lyndon Evans (Lausanne: EPFL Press, 2009), 23). Wilczek discusses the relation between mass, matter, and embodiment as follows: "The mass of ordinary matter is the embodied energy of more basic building blocks, themselves lacking mass" (*The Lightness of Being*, 2).

153. See Ellis, "The Fundamental Physics Behind the LHC."

154. See Jacques Derrida and Bernard Stiegler, *Echographies of Television* (Cambridge: Polity Press, 2002).

155. Affect here is understood as that which is beyond or outside of social signification (see Clare Hemmings, "Invoking Affect: Cultural Theory and the Ontological Turn," *Cultural Studies* 19, no. 5 (2005): 548–567) and as a physical and emotional tendency (see Hardt, "Foreword"). Though useful in these respects, the concept is problematic for us partly in that it offers up a rather mechanistic model of reciprocity or mutuality. As Hardt puts it, "Affects require us, as the term suggests, to enter the realm of causality . . . they illuminate . . . both our power to affect the world around us and our power to be affected by it" ("Foreword," ix).

156. Barad, *Meeting the Universe Halfway*, 35.

157. Merrin, *Baudrillard and the Media*, 63.

158. The United Kingdom is reported to have contributed 70 percent of the funding for the LHC project.

159. Matthew Chalmers, "CERN: The View from Inside," http://physicsworld.com, January 27 (2009).

160. Chalmers, "CERN," 2.

161. Brian Cox, quoted in *The Observer*, March 7, 2010.

162. John Ellis, Gian Giudice, Michelangela Mangano, Igor Tkachev, and Urs Wiedemann (the LHC Safety Assessment Group). "Review of the Safety of LHC Collisions." LHC Safety Assessment Group, accessed July 7, 2011, http://cern.ch/lsag/LSAG-Report.pdf.

163. *Angels and Demons*, the "Dan Brown novel in which antimatter is stolen from CERN to destroy the Vatican"; Chalmers, "CERN," 2.

164. James Dacey, "Breathe Easy, the LHC Still Won't Swallow the Earth," accessed March 1, 2010, http://physicsworld.com/blog/2010/02/breathe_easy_the_lhc_still_won.htm.

165. YouTube Countdown CERN—BBC News.

166. See http://www.bbc.co.uk/radio4/bigbang.

167. See Christine McGourty, *BBC NEWS* (September 10, 2008).

168. BBC Four, "The Big Bang Machine" (September 4, 2008).

169. Baudrillard, *Simulations*.

170. Merrin, *Baudrillard and the Media*, 68.

171. Barad, *Meeting the Universe*, 46.

172. Alfred Gell, "The Technology of Enchantment and the Enchantment of Technology," in *Anthropology, Art and Aesthetics*, ed. Jeremy Coote and Anthony Shelton (Oxford: Clarendon, 1992), 40–66.

173. Lucy Suchman, *Human–Machine Reconfigurations* (Cambridge: Cambridge University Press, 2007), 244.

174. Morris, Hazel, "List of Facts for the PPARC website," accessed July 7, 2011, http://lhc-machine-outreach.web.cern.ch/lhc-machine-outreach/lhc-interesting-facts.htm.

175. LHC Machine Outreach, accessed July 7, 2011, http://lhc-machine-outreach.web.cern.ch.

176. Barad, *Meeting the Universe Halfway*, 49.

177. Fenton O'Creavy offers this definition of performativity in the context of debates on the credit crunch: "performative statements or beliefs are those which help bring about the conditions they describe" ("Has Robert Peston Caused a Recession?"). Similarly, Callon writes: "scientific theories, models, and statements are not constative; they are performative, that is, actively engaged in the constitution of the reality that they describe" ("What Does It Mean to Say That Economics Is Performative?," 318).

178. Barad, *Meeting the Universe Halfway*, 49.

179. Barad, discussing the work of Ian Hacking, reminds us that microscopic entities are always in part artifacts of experimentation and visualization technologies. A scanning tunneling microscope, for example, "undermines any illusion that the image represents the mere magnification of what we see with our eyes" (*Meeting the Universe*, 51).

180. Elizabeth Grosz, *The Nick of Time: Politics, Evolution and the Untimely* (Durham, NC: Duke University Press, 2004), 239.

181. Merrin, *Baudrillard and the Media*, 71.

182. Merrin, *Baudrillard and the Media*, 63.

183. Baudrillard, *Simulations*, 55.

184. Baudrillard, *The Transparency of Evil*, 36.

185. In these contexts: texts, tweets, blogs, photographs, videos, and so on.

186. See *Wonders of the Solar System*, BBC 2, March 2010.

187. BBC Four, 2008.

188. Ellis, "The Fundamental Physics Behind the LHC," 33.

189. Worldwide LHC Computing Grid Technical Site, accessed July 7, 2011, http://lcg.web.cern.ch.

190. GridCafé, accessed July 7, 2011, http://www.gridcafe.org.

191. See Ian Foster and Carl Kesselman (eds.), *The Grid: Blueprint for a New Computing Infrastructure* (San Francisco: Morgan Kaufmann, 2004).

192. GridCafé.

193. Open Grid Forum, accessed July 7, 2011, http://www.ogf.org.

194. Pickering argues that constructionism is essentially humanist and cannot take account of material agency understood as "agency that comes at us from outside the human realm and that cannot be reduced to anything within that realm" (*The Mangle of Practice*, 6).

195. This is the case whatever we know of the five-second delay in the acquisition of data from the detectors, because during those five seconds "nobody knows quite what is happening." The event happens onscreen.

196. Durkheim, *The Elementary Forms of the Religious Life*, 53.

197. Baudrillard, *Transparency of Evil*, 36.

198. In as far as intuition is a sense, it should be distinguished from what Bergson referred to in *Creative Evolution* as the bodily senses, including touch. Baudrillard reminds us that it was McLuhan who foresaw an "era of *tactile* communication" (*Simulations*, 123).

Chapter 3

1. The argument about the special relation between photography and the practice of the cut forms the core of this chapter. Put briefly, the reasons for this "specialness" (which is not the same things as exclusivity) are multiple, the predominant one having to do with the way in which, as we suggest later, photography highlights rather than hides the process of cutting. It does this both on an ontological level (by carving reality into fragments and framing them for us in a certain way) and on an ethical level (whereby ethical stakes are raised by its supposedly evidential force that makes a claim to a certain version of reality and demands a response from us to it).

2. Susan Sontag, "Regarding The Torture of Others," *New York Times*, May 23, 2004. Retrieved July 14, 2011, from http://www.nytimes.com/2004/05/23/magazine/regarding-the-torture-of-others.html.

3. http://www.thehighline.org/galleries/images/joel-sternfeld.

4. http://www.richardgalpin.co.uk/introduction.php.

5. Gilles Deleuze and Félix Guattari, *What Is Philosophy?*, trans. Hugh Tomlinson and Graham Burchell (New York: Columbia University Press,1994), 208.

6. Eugene Holland, "Concepts and Utopia," *The Deleuze Dictionary*, ed. Adrian Parr (Edinburgh: Edinburgh University Press, 2005), 52.

7. Deleuze and Guattari, *What Is Philosophy?*, 16.

8. Hollis Frampton, "Some Propositions on Photography," *On the Camera Arts and Consecutive Matters: The Writings of Hollis Frampton*, ed. Bruce Jenkins (Cambridge, MA: MIT Press, 2009), 8.

9. David Bate, *Photography: Key Concepts* (Berg: Oxford, 2009), 2.

10. Sarah Kember, "The Virtual Life of Photography," *Photographies* 1, no. 2 (September 2008): 177.

11. Kember, "The Virtual Life of Photography," 181.

12. Henri Bergson, *Creative Evolution*, trans. Arthur Mitchell (New York: The Modern Library, 1944), 169.

13. Bergson, *Creative Evolution*, 178.

14. Bergson, *Creative Evolution*, 181.

15. Bergson, *Creative Evolution*, 182.

16. Bergson, *Creative Evolution*, 182.

17. Bergson, *Creative Evolution*, 179.

18. Bergson, *Creative* Evolution, 362.

19. Gilles Deleuze, *Cinema 1: The Movement-Image*, trans. Hugh Tomlinson and Barbara Habberjam (London: The Athlone Press, 1986), 24.

20. Alexander Sekatskiy, "The Photographic Argument of Philosophy," *Philosophy of Photography* 1, no. 1 (2010): 83.

21. Bergson, *Creative Evolution*, 223.

22. Bergson, *Creative Evolution*, 248–249.

23. Bergson, *Creative Evolution*, 272.

24. Gilles Deleuze, *Bergsonism*, trans. Hugh Tomlinson and Barbara Habberjam (New York: Zone Books, 1988), 25.

25. Deleuze, *Bergsonism*, 17.

26. Deleuze, *Bergsonism*, 35.

27. See Bergson, *Creative Evolution*, 335.

28. Hollis Frampton, "Eadweard Muybridge: Fragments of a Tesseract," *On the Camera Arts and Consecutive Matters*, 30, emphasis in original.

29. Jacques Derrida and Bernard Stiegler, *Echographies of Television*, trans. Jennifer Bajorek (Cambridge: Polity Press, 2002), 76, emphasis in original.

30. Bergson, *Creative Evolution*, 4.

31. "*No doubt, it is useful to us*, in view of our ulterior manipulation, to regard each object as divisible into parts arbitrarily cut up, each part being again divisible as we like, and so on ad infinitum," Bergson, *Creative Evolution*, 169–170, emphasis added.

32. Karen Barad, *Meeting the Universe Halfway: Quantum Physics and the Entanglement of Matter and Meaning* (Durham, NC: Duke University Press, 2006), 62–64.

33. Arguably, this is an addition that comes at a price, which is evident in Barad's somewhat perplexing, and of course language-based, declaration that "Language has been granted too much power," *Meeting the Universe Halfway*, 132.

34. Barad, *Meeting the Universe Halfway*, 43.

35. Barad, *Meeting the Universe Halfway*, 40.

36. Barad, *Meeting the Universe* Halfway, 142, emphasis in original.

37. Barad, *Meeting the Universe Halfway*, 142.

38. Barad, *Meeting the Universe Halfway*, 140.

39. Barad, *Meeting the Universe Halfway*, 145.

40. Barad, *Meeting the Universe Halfway*, 142.

41. Deleuze, *Bergsonism*, 94.

42. Claire Colebrook, *Deleuze and the Meaning of Life* (London and New York: Continuum 2010), 11.

43. Kember, "The Virtual Life of Photography," 182.

44. Bernard Stiegler, "The Discrete Image," in Jacques Derrida and Bernard Stiegler, *Echographies of Television*, trans. Jennifer Bajorek (Cambridge: Polity Press, 2002, 147, emphasis in original.

45. Stiegler, "The Discrete Image," 147.

46. As Colebrook puts it in *Deleuze and the Meaning of Life*:

The intellect comes to regard itself as a quantifiable and manipulable entity, and therefore imagines its perception of the world as a series of snapshots that it must tie together in order to form a continuous reality. Against this idea of mind as cinematic machine, Bergson will insist upon a perception which, prior to the bifurcation between intellect and instinct, had not yet determined the world in advance through some proto-machinic technology of fixed concepts, discrete units and comparable quantities. Indeed, in order for mind to discover its proper power, it needs to return, in a vitalist manner, to the very power *to quantify, to conceptualize or to reify* that now impedes its further creative evolution. (9)

47. Stiegler, "The Discrete Image," 148.

48. See Deleuze, *Bergsonism*, 51.

49. Stiegler, "The Discrete Image," 148.

50. Stiegler, "The Discrete Image," 152.

51. Stiegler, "The Discrete Image," 153, emphasis in original.

52. Stiegler, "The Discrete Image," 162.

53. Stiegler, "The Discrete Image," 162.

54. Sekatskiy, "The Photographic Argument of Philosophy," 85.

55. Sekatskiy, "The Photographic Argument of Philosophy," 85.

56. Jonathan Friday, "Stillness Becoming: Reflections on Bazin, Barthes, and Photographic Stillness," in *Stillness and Time: Photography and the Moving Image*, ed. David Green and Joanna Lowry (Brighton, UK: Photoworks, 2006), 42.

57. Joanna Zylinska, "On Bad Archives, Unruly Snappers and Liquid Photographs," *Photographies* 3, no. 2 (August 2010): 150.

58. Some of the material on *Oblique* enclosed in this chapter was originally published as a catalog essay, "The Cut of the Artist: Sellars' Anatomy Lesson," written by Joanna Zylinska for Nina Sellars's exhibition at Guildford Lane Gallery in Melbourne in 2008.

59. Bergson, *Creative Evolution*, 15.

60. The *Oblique* show was first presented at Guildford Lane Gallery in Melbourne in August 2008.

61. Hollis Frampton, "Digressions on the Photographic Agony," *On the Camera Arts and Consecutive Matters*, 21.

62. See Victor Burgin, *Components of a Practice* (Milan: Skira, 2008).

63. Frampton, "Digressions on the Photographic Agony," 21, emphasis in original.

64. Hent De Vries, "Violence and Testimony: On Sacrificing Sacrifice," in *Violence, Identity and Self-Determination*, ed. Hent De Vries and Samuel Weber (Stanford: Stanford University Press, 1997), 29. See also Emmanuel Levinas, *Totality and Infinity: An Essay on Exteriority*, trans. Alphonso Lingis (Pittsburgh: Duquesne University Press, 1969), 194–197.

65. Joanna Zylinska, *The Ethics of Cultural Studies* (London and New York: Continuum, 2002), 74–75.

66. I am grateful to Mary Hunter for her helpful comments on seventeenth-century anatomy lessons.

67. Gunther von Hagens's televised autopsies, shown as part of *Anatomy for Beginners* in 2005 and of *Autopsy: Life and* Death in 2006 on the United Kingdom–based Channel 4, are examples of such programs.

68. Slavoj Žižek, *The Ticklish Subject: The Absent Centre of Political Ontology* (London: Verso, 1999), 37–71.

69. Ron Burnett, *How Images Think* (Cambridge, MA: MIT Press, 2004), 7.

70. Parveen Adams, *The Emptiness of the Image: Psychoanalysis and Sexual Difference* (London: Routledge, 1996), 145.

71. Jay David Bolter and Richard Grusin, *Remediation: Understanding New Media* (Cambridge, MA: MIT Press, 2000), 11.

72. Victor Burgin, "Looking at Photographs," *Thinking Photography* (London: Macmillan, 1982), 152.

Chapter 4

1. Henri Bergson, *Creative Evolution*, trans. Arthur Mitchell (New York: The Modern Library, 1944), xxv.

2. We are referring here to Carolyn Marvin, *When Old Technologies Were New* (New York: Oxford University Press, 1988), also discussed in chapter 1, as an example of such thinking.

3. We are referring here to developments in cybernetics, Artificial Intelligence, Artificial Life, and genetic engineering.

4. In *Transpositions* (Cambridge: Polity Press, 2006), Rosi Braidotti similarly speaks of renewed master narratives such as biotechnology and globalization that seem to be all about change but are in fact anything but.

5. See Bergson, *Creative Evolution*; Rosi Braidotti, *Metamorphoses* (Cambridge: Polity Press, 2002).

6. See Joshua S. Hanan, "Home Is Where the Capital Is: The Culture of Real Estate in an Era of Control Societies," *Communication and Critical/Cultural Studies* 7, no. 2 (2010): 176–201.

7. The "data subject" is also a legal term, standing for an individual who is the subject of personal information (data).

8. See Mark Weiser, "The Computer for the 21st Century," *Scientific American* 265, September (1991): 94–104.

9. See Wim F. J. Verhaegh, Emile Aarts, and Jan Korst (eds.), *Algorithms in Ambient Intelligence* (Boston: Kluwer Academic Publishers, 2004).

10. Emile Aarts, Jan Korst, and Wim F. J. Verhaegh, "Algorithms in Ambient Intelligence" (2004) in Verhaegh, Aarts, and Korst, *Algorithms in Ambient Intelligence*, 6.

11. See Adam Greenfield, *Everyware: The Dawning Age of Ubiquitous Computing* (Berkeley: New Riders Publishing, 2006).

12. Such criteria include: self-organization, self-replication, evolution, autonomy, and emergence. See Christopher G. Langton, "Artificial Life," in *The Philosophy of Artificial Life*, ed. Margaret A. Boden (Oxford and New York: Oxford University Press, 1996).

13. Stefano Marzano, author of "Cultural Issues in Ambient Intelligence" (in *The New Everyday: Views on Ambient Intelligence*, ed. Stefano Marzano and Emile Aarts [Rotterdam: 010 Publishers, 2003]) regards Ambient Intelligence as a naturalized form of Artificial Intelligence: "We may find that any non-interactive objects or systems around us have been replaced by almost invisible, intelligent information systems—an 'AI' that could soon form a natural part of our everyday lives" (8).

14. Aarts, Korst, and Verhaegh, "Algorithms in Ambient Intelligence," 4; Greenfield, *Everyware*, 3.

15. Marzano, "Cultural Issues in Ambient Intelligence," 9.

16. Fiona Allon, "An Ontology of Everyday Control: Space, Media Flows and 'Smart' Living in the Absolute Present," in *Media Space: Place, Scale and Culture in a Media Age*, ed. Nick Couldry and Anna McCarthy (London: Routledge, 2003), 258.

17. See "House of the Future," parts 1 and 2, http://www.youtube.com/watch?v=DoCCO3GKqWY and http://www.youtube.com/watch?v=lVMAeSNZZz0.

18. See Frederic P. Miller, Agnes F. Vandome, and John McBrewster (eds.), *Monsanto* (Beau Bassin, Mauritius: Alphascript Publishing, 2009).

19. See "House of the Future," parts 1 and 2.

20. Inside Microsoft's "FutureHome," http://www.youtube.com/watch?v=ODpRCOKQUXM.

21. Danny Briere and Pat Hurley, *Smart Homes for Dummies* (Hoboken, NJ: Wiley Publishing, Inc., 2007), 12.

22. Briere and Hurley, *Smart Homes*, 13.

23. Briere and Hurley, *Smart Homes*, 13.

24. Briere and Hurley, *Smart Homes*, 14.

25. Briere and Hurley, *Smart Homes*, 15.

26. Briere and Hurley, *Smart Homes*, 15.

27. Betty Friedan, *The Feminine Mystique* (London: Penguin Books, 2010), 190.

28. Friedan, *The Feminine Mystique*, 195.

29. In Friedan, *The Feminine Mystique*, viii.

30. See Angela McRobbie, *The Aftermath of Feminism: Gender, Culture and Social Change* (London: Sage, 2009).

31. In her *Metamorphoses* and *Transpositions*, Braidotti has commented on the paucity of the imagination in technoscientific cultures that otherwise promise, in part through technoscientific innovation, to push the imagination to its limits. Our point is that even if the recurrence of gender stereotypes is a facet of this failure of imagination, it is still—and always—significant.

32. Marzano, "Cultural Issues in Ambient Intelligence," 8.

33. See, for example, Alison Adam, *Artificial Knowing: Gender and the Thinking Machine* (London: Routledge, 1998); Paul N. Edwards, *The Closed World: Computers and the Politics of Discourse in Cold War America* (Cambridge, MA: MIT Press, 1996); N. Katherine Hayles, *How We Became Posthuman: Virtual Bodies in Cybernetics, Literature and Informatics* (Chicago: University of Chicago Press, 1999); Stefan Helmreich, *Silicon Second Nature: Culturing Artificial Life in a Digital World* (Berkeley: University of California Press, 1998); Sarah Kember, *Cyberfeminism and Artificial Life* (London: Routledge, 2003); Lucy Suchman, *Human–Machine Reconfigurations* (Cambridge: Cambridge University Press, 2007).

34. Arthur C. Clarke's *2001: A Space Odyssey* (London: Orbit, 2000) was originally published in 1968, after the release of Kubrick's film. In the film, Hal 9000 and Dave Bowman confront each other about reentry into the spaceship, whereas in the novel they are in dispute over access, or otherwise, to manual control:

> "I want to do this myself, Hal," he said. "Please give me control."
> "Look, Dave, you've got a lot of things to do. I suggest you leave this to me." (158)

In both cases, the scene represents a turning point, which is encapsulated not just by the acquisition of control from computer to human but also by a kind of switching of the sides, or roles, whereby we see, through the disconnection of Hal, "his" humanity (expressed as and through the display of emotion) and Dave's roboticism (expressed through his logical and unemotive execution of a difficult task).

35. Quotations are from Stanley Kubrick's 1968 film *2001: A Space Odyssey*.

36. Brian Aldiss, *Supertoys Last All Summer Long and Other Stories of Future Time* (London: Orbit, 2001).

37. Douglas Adams, *The Hitch-Hiker's Guide to the Galaxy* (London: Pan Books Ltd, 1981), 73.

38. Adams, *The Hitch-Hiker's Guide*, 75.

39. See David G. Stork (ed.), *Hal's Legacy: 2001's Computer as Dream and Reality* (Cambridge, MA: MIT Press, 1996).

40. Philip K. Dick, *UBIK* (London: Orion, 2004), 28.

41. Mark Andrejevic, *iSpy: Surveillance and Power in the Interactive Era* (Lawrence: University Press of Kansas, 2007), 122.

42. Greenfield, *Everyware*, 49.

43. Greenfield, *Everyware*, 49.

44. Suchman, *Human–Machine Reconfigurations*, 213.

45. Suchman, *Human–Machine Reconfigurations*, 213.

46. See Lin Padgham and Michael Winikoff, *Developing Intelligent Agent Systems* (Chichester, UK: John Wiley and Sons Ltd., 2004).

47. See Hayles, *How We Became Posthuman*; Kember, *Cyberfeminism and Artificial Life*; Suchman, *Human–Machine Reconfigurations*.

48. For a discussion of the latter issue, see Chris Hables Gray, *Cyborg Citizen* (New York: Routledge, 2002).

49. John Cass, Lorna Goulden, and Slava Kozlov, "Intimate Media: Emotional Needs and Ambient Intelligence," in Marzano and Aarts, *The New Everyday*, 219.

50. Cass, Goulden, and Kozlov, "Intimate Media," 220.

51. Cass, Goulden, and Kozlov, "Intimate Media," 220.

52. Marzano, "Cultural Issues in Ambient Intelligence," 9.

53. Suchman, *Human–Machine Reconfigurations*, 220.

54. Marzano, "Cultural Issues in Ambient Intelligence," 9.

55. Greenfield, *Everyware*, 81.

56. Greenfield, *Everyware*, 81–82.

57. Andrejevic, *iSpy*, 132.

58. Andrejevic, *iSpy*, 132.

59. Suchman, *Human–Machine Reconfigurations*, 268.

60. Suchman, *Human–Machine Reconfigurations*, 269.

61. Aarts, Korst, and Verhaegh, "Algorithms in Ambient Intelligence," 5.

62. See Rosalind Picard, "Does HAL Cry Digital Tears? Emotion and Computers," in Stork, *Hal's Legacy*.

63. For an analysis of this relationship, see Kember, *Cyberfeminism and Artificial Life*.

64. Or it regards one as the expression of the other. See Suchman, *Human–Machine Reconfigurations*, 233.

65. Michael Hardt, "Foreword: What Affects are Good For?," in *The Affective Turn: Theorizing the Social*, ed. Patricia Ticineto Clough with Jean Halley (Durham: Duke University Press, 2007), xi.

66. Hardt, "Foreword," xi.

67. Hardt, "Foreword," xii.

68. Hardt, "Foreword," iv.

69. See Karen Barad, *Meeting the Universe Halfway* (Durham, NC: Duke University Press, 2007).

70. See Suchman, *Human–Machine Reconfigurations*.

71. The difference we are alluding to is between a spatial model of mutuality and a temporal model of relationality. Rather than suggesting that one is inherently better than the other, we are arguing that where the former is more circular and self-reinforcing, the latter is more transformative.

72. In "Invoking Affect: Cultural Theory and the Ontological Turn" (*Cultural Studies* 19, no. 5 (2005): 548–567), Clare Hemmings also refers to the "self-referential" (552) characteristic of affects as "states of being" (551) that may manifest or be interpreted as emotions. She cites the work of psychologist Silvan Tomkins, who "asked us to think of the contagious nature of a yawn, smile or blush. It is transferred to others and doubles back, increasing its original intensity" (552). If affect itself is cyclical, then, for Hemmings, there is also something inherently circular about the so-called affective turn in cultural theory—a turn that "flattens out," or excises, the bodily, nonrational, and potentially transformative states of being from earlier poststructuralist theories in order to (re)turn to them anew. It is significant for our argument here that one of the theories so flattened in the so-called affective turn, according to Hemmings, is feminist standpoint epistemology.

73. See Abraham Maslow, *Motivation and Personality* (New York: Longman, 1987).

74. Cass, Goulden, and Kozlov, "Intimate Media," 218.

75. Cass, Goulden, and Kozlov, "Intimate Media," 218.

76. Cass, Goulden, and Kozlov, "Intimate Media," 218.

77. Greenfield, *Everyware*, 3.

78. Greenfield, *Everyware*, 3.

79. Andrejevic, *iSpy*, 95.

80. Andrejevic, *iSpy*, 94.

81. Andrejevic, *iSpy*, 96.

82. Andrejevic, *iSpy*, 97.

83. Andrejevic, *iSpy*, 97, 98.

84. Allon, "An Ontology of Everyday Control," 264.

85. Allon, "An Ontology of Everyday Control," 268.

86. Sarah Kember, "Doing Technoscience as (New) Media," in *Media and Cultural Theory*, ed. James Curran and David Morley (London: Routledge, 2004), 241.

87. Gary Hall, *Digitize This Book! The Politics of New Media, or Why We Need Open Access Now* (Minneapolis: University of Minnesota Press, 2008), 15.

88. Hall, *Digitize This Book!*, 17.

89. It does so in a way that ALife did not, even though it reduced life to a set of computational criteria.

90. Darin Barney, "One Nation Under Google: Citizenship in the Technological Republic," The 2007 Hart House Lecture. Hart House Lectures/Coach House Books. See also Barney's "Radical Citizenship in the Republic of Technology: A Sketch," in *Radical Democracy and the Internet*, ed. Lincoln Dahlberg and Eugenia Siapera (New York: Palgrave Macmillan, 2007), 37–54.

91. Edwards, *The Closed World*, 1.

92. Edwards, *The Closed World*, 1.

93. Edwards, *The Closed World*, 2.

94. Edwards, *The Closed World*, 3.

95. Edwards, *The Closed World*, 3.

96. This idea was put forward by Steve Grand in his *Creation: Life and How to Make It* (London: Weidenfeld & Nicolson, 2000).

97. Kember, *Cyberfeminism and Artificial Life*, 208. ALife's feminism was arguably "a kind of essentialist eco-feminism—a nostalgic 1960s earth mother to the industrialist-militarist patriarchal cyborg figure of the 1980s" (208). Haraway's (1991) cyborg politics parodied the macho robots of the Cold War, and if she reminds us that there is nothing "soft" about biology, we could also argue that there is nothing inherent in the politics it evokes, or in "the displacement (which, with the inauguration of US President George W. Bush and the events of 11 September 2001 is likely to be temporary) of the arms race by ideologies of greater co-operation" (208).

98. David Lyon, *Surveillance after September 11* (Cambridge: Polity, 2008), 4.

99. Lyon, *Surveillance after September 11*, 5.

100. Lyon, *Surveillance after September 11*, 9.

101. Lyon, *Surveillance after September 11*, 10.

102. Wendy Brown, *Edgework: Critical Essays on Knowledge and Politics* (Princeton: Princeton University Press, 2005), 47.

103. Brown, *Edgework*, 47.

104. Brown, *Edgework*, 37.

105. Brown, *Edgework*, 40.

106. Brown, *Edgework*, 43.
107. Braidotti, *Transpositions*, 1.
108. Brown, *Edgework*, 40.
109. Brown, *Edgework*, 42.
110. Brown, *Edgework*, 42.
111. Andrejevic, *iSpy*, 106.
112. "Who's Watching You?," *BBC 2,* May 25—June 8, 2009.
113. Won on appeal.
114. Andrejevic, *iSpy*, 111.
115. Edwards, *The Closed World*, 3.
116. See David Lyon, *Surveillance as Social Sorting* (London: Routledge, 2003).
117. Andrejevic, *iSpy*, 124.
118. David Morley, *Home Territories: Media, Mobility and Identity* (London: Routledge, 2000), 64.
119. Morley, *Home Territories*, 97.
120. Morley, *Home Territories*, 78.
121. Haraway cited in Morley, *Home Territories*, 77.
122. Andrejevic, *iSpy*, 125.
123. Lyon, *Surveillance after September 11*, 88.
124. Lyon, *Surveillance after September 11*, 50.
125. Lyon, *Surveillance after September 11*, 99.
126. Garland in Lyon, *Surveillance after September 11*, 102.
127. Brown, *Edgework*, 48.
128. Lyon, *Surveillance after September 11*, 102.
129. See Andrejevic's *iSpy* for others, including "portable networked devices, cars, and cell phones that know where they are" (111).
130. See Joseph Turow, *Niche Envy* (Cambridge, MA: MIT Press, 2008).
131. In Morley, *Home Territories*, 149.
132. See Morley, *Home Territories*.
133. Andrejevic, *iSpy*, 113.
134. In Andrejevic, *iSpy*, 113.
135. Andrejevic, *iSpy*, 114.
136. Kember, *Cyberfeminism and Artificial Life*, 6.
137. Greenfield, *Everyware*, 180.
138. Greenfield, *Everyware*, 180.
139. Greenfield, *Everyware*, 246.
140. Greenfield, *Everyware*, 148.
141. Lyon, *Surveillance as Social Sorting*, 27.
142. Lyon, *Surveillance as Social Sorting*, 20.
143. Brown, *Edgework*, 59.
144. Brown, *Edgework*, 59.
145. As such, the self can of course be regarded as discriminatory and exclusive: a result of social sorting. Challenges to this concept have come not only from feminism but equally from other fields such as poststructuralism, postcolonial studies, and queer theory.

146. Bergson, *Creative Evolution*, 7, 25, 56.

147. Suchman, *Human–Machine Reconfigurations*, 275.

148. Suchman, *Human–Machine Reconfigurations*, 283.

149. Dick, *UBIK*, 23.

150. Dick, *UBIK*, 81.

151. Dick, *UBIK*, 178.

152. We are of course aware of the technicist tradition of science fiction and its tendency toward a limited dualism of utopias and dystopias. In other words, we are not suggesting that the "answer" lies in science fiction per se or, indeed, in a turn from theory to practice.

153. Hall, *Digitize This Book!*, 29.

154. In her article, "Media, Mars, Metamorphosis" (*Culture Machine* 11, 2010, accessed July 7, 2011, http://www.culturemachine.net)—subsequently rewritten by both of us as an epigraph for this book—Kember has reported on a counterindustry experiment or, rather, on an industry experiment that might be deemed to have gone wrong. It was designed not by employees at Phillips but by a Swedish team of researchers who thought it would be interesting to see what happened if they put a Luddite in a smart home and did not let him out for eight weeks. This particular Luddite (they called him Hal) was a middle-aged, unemployed (which is why, for a small fee, he agreed to take part), divorced ex-alcoholic. He was not good at taking care of himself or others and did not appreciate the advances of electronic artifacts that wanted, for whatever reason, to get to know him. His response lacked what Derrida would call a certain hospitality. Indeed, it bordered on outright hostility. He refused the tasks that were assigned to him (shopping online, joining a social networking site, creating his own webpage, and engaging in conversation with a speech-based autonomous agent called Dave) and, due to a combination of boredom, extreme irritation, and an increased—almost existential—sense of loneliness, he walked out of the house before his eight weeks were up. Because there was no light switch, he did not turn out the lights (he had never figured out how) and the cameras in his ceiling that were recording the experiment kept filming the empty room and the array of objects that hummed without a tune, relating the details of his life among themselves. When, some time later, he looked at the film on YouTube, Hal saw, or thought he saw, more than a trace or a spark of his life in those objects. He communicated this to Jeremy Hoyle, a cultural theorist and disciple of Francis Fukuyama, who was writing a book about him and others involved in current biotech experiments. Hoyle did not believe Hal—at least, not until his mirror and his toilet started talking to him and he could not figure out how to stop them.

155. See Joanna Zylinska, *Bioethics in the Age of New Media* (Cambridge, MA: MIT Press, 2009), 30.

Chapter 5

1. Rosi Braidotti, *Transpositions: On Nomadic Ethics* (Cambridge: Polity Press, 2006), 1.

2. Elizabeth Grosz, "Becoming . . . An Introduction," in *Becomings: Explorations in Time, Memory, and Futures*, ed. Elizabeth Grosz (Ithaca: Cornell University Press, 1999), 4.

3. Henri Bergson, *Creative Evolution*, trans. Arthur Mitchell (New York: The Modern Library, 1944), 374.

4. Bergson, *Creative Evolution*, 296.

5. See Braidotti, *Transpositions*.

6. Bernadette Wegenstein, *Getting Under the Skin: The Body and Media Theory* (Cambridge, MA: MIT Press, 2006), 3.

7. Jay David Bolter and Richard Grusin, *Remediation: Understanding New Media*, Cambridge, MA: MIT Press, 2000), 232.

8. Bolter and Grusin, *Remediation*, 232.

9. Bolter and Grusin, *Remediation*, 233.

10. Bolter and Grusin, *Remediation*, 233.

11. Bolter and Grusin, *Remediation*, 233.

12. Bolter and Grusin, *Remediation*, 236.

13. Interestingly, in his *Premediation: Affect and Mediality After 9/11* (Basingstoke, UK: Palgrave Macmillan, 2010), Richard Grusin challenges Mark Hansen precisely for such a naive reading of the possibility of "empowering" the individual and challenging the hegemony of the culture industry "by facilitating personal control over the flux of time—whether this be the flux of television in one's living room (think of the potential of TIVO and other digital storage systems)" (Hansen, quoted in Grusin, 36). Indeed, Grusin raises some serious doubt as to the extent to which such consumer-driven behavior can amount to "cultural resistance."

14. Bolter and Grusin, *Remediation*, 58.

15. Bolter and Grusin, *Remediation*, 59.

16. Bolter and Grusin, *Remediation*, 238.

17. Virginia L. Blum argues in *Flesh Wounds: The Culture of Cosmetic Surgery* (Berkeley: University of California Press, 2003) that Ovid's myth of *Pygmalion* is reworked in cosmetic surgery so that the surgeon/sculptor becomes intrigued with the object of his creation, in this case a flesh rather than stone idealization of the female body.

18. Bolter and Grusin, *Remediation*, 238.

19. Michel Foucault, "Technologies of the Self," in *Technologies of the Self*, ed. Luther H. Martin, Huck Gutman, and Patrick H. Hutton (London: Tavistock Books, 1988), 18.

20. Michel Foucault, *The Birth of Biopolitics: Lectures at the College de France, 1978–79*, trans. Graham Burchell (Basingstoke, UK: Palgrave Macmillan, 2008), 295.

21. This definition corresponds to the first meaning of the term "governmentality," which Foucault outlines in his lecture from February 1, 1978, included in *Security, Territory, Population: Lectures at the Collège de France 1977–1978* (Basingstoke, UK: Palgrave Macmillan, 2007). He explains there:

> By this word "governmentality" I mean three things. First, by "governmentality" I understand the ensemble formed by institutions, procedures, analyses and reflections, calculations, and tactics that allow the exercise of this very specific, albeit very complex, power that has the population as its target, political economy as its major form of knowledge, and apparatuses of security as its essential technical instrument. Second, by "governmentality" I understand the tendency, the line of force, that for a long time, and throughout the West, has constantly led toward the pre-eminence over all other types of power—sovereignty, discipline, and so on—of the type of power that we can call "government" and which has led to the development of a series of specific governmental apparatuses (*appareils*) on the one hand, [and, on the other] to the development of a series of knowledges (*savoirs*). Finally, by "governmentality" I think we should understand the process, or rather, the result of the process by which the state of justice of the Middle Ages became the administrative state in the fifteenth and sixteenth centuries and was gradually "governmentalized." (108–109)

22. Mildred Blaxter, "The Case of the Vanishing Patient? Image and Experience," *Sociology of Health and Illness* 31, no. 5 (2009), 1.

23. Blaxter, "The Case of the Vanishing Patient?," 1.

24. Blum, *Flesh Wounds*, 101.

25. Quoted in Blaxter, "The Case of the Vanishing Patient?," 2.

26. Quoted in Blaxter, "The Case of the Vanishing Patient?," 3.

27. Blaxter, "The Case of the Vanishing Patient?," 10.

28. Blaxter, "The Case of the Vanishing Patient?," 13.

29. Blaxter, "The Case of the Vanishing Patient?," 15.

30. Susan Sontag, "Regarding the Torture of Others," *New York Times* (May 23, 2004), accessed July 14, 2011, http://www.nytimes.com/2004/05/23/magazine/regarding-the-torture-of-others.html.

31. Sontag, "Regarding the Torture of Others."

32. Grusin develops similar argument in chapter 3 of his *Premediation*. He writes, "Part of the force of the Abu Ghraib photographs comes precisely from their participation in our technological nonconscious—the way in which they are integrated within our everyday unconscious use of technology" (72).

33. Blum, *Flesh Wounds*, 101.

34. See Karen Barad, *Meeting the Universe Halfway* (Durham, NC: Duke University Press, 2007), 50–51.

35. Blum, *Flesh Wounds*, 156.

36. See Christopher Lasch, *The Culture of Narcissism* (London: Abacus, 1979).

37. See Meredith Jones, *Skintight: An Anatomy of Cosmetic Surgery* (Oxford: Berg, 2008).

38. Jones, *Skintight*, 3.

39. Jones, *Skintight*, 1.

40. Quoted in Blum, *Flesh Wounds*, 73.

41. Quoted in Blum, *Flesh Wounds*, 73.

42. Victoria Pitts-Taylor, *Surgery Junkies: Wellness and Pathology in Cosmetic Surgery* (New Brunswick, NJ: Rutgers University Press, 2003), 3.

43. Blum, *Flesh Wounds*, 271.

44. Jones, *Skintight*, 53.

45. Joanna Zylinska, *Bioethics in the Age of New Media* (Cambridge, MA: MIT Press, 2009), 114.

46. Wegenstein, *Getting Under the Skin*, 6.

47. Jones, *Skintight*, 107.

48. Blum, *Flesh Wounds*, 146.

49. Blum, *Flesh Wounds*, 146.

50. Blum, *Flesh Wounds*, 229.

51. Blum, *Flesh Wounds*, 223.

52. See Jones, *Skintight*.

53. Pitts-Taylor, *Surgery Junkies*, 2.

54. Pitts-Taylor, *Surgery Junkies*, 10.

55. Pitts-Taylor, *Surgery Junkies*, 26.

56. Foucault, "Technologies of the Self," 19.

57. Foucault, "Technologies of the Self," 18.

58. Michel Foucault, *The Hermeneutics of the Subject: Lectures at the Collège de France 1981–82*, ed. Frederic Gros and Francois Ewald (Basingstoke, UK: Palgrave Macmillan, 2005), 8.

59. Michel Foucault, *Ethics: Subjectivity and Truth*, ed. Paul Rabinow (London: Allen Lane, 1997), 269–271. See Zylinska, *Bioethics in the Age of New Media*, 77. Also, chapter 4 in Zylinska's *Bioethics* offers a reading of makeover culture in the context of biopolitics.

60. In the process, "As consumers, we scout the Web for the latest health research, preventative measures, medical trials, and cosmetic procedures. We assume greater responsibilities in our exchanges with doctors and clinicians. We take on a wider array of body projects related to health, from dieting to yoga. We take more elective medicines, from those to improve our sex lives to those that limit menstruation, or help us sleep, concentrate, relax, perform athletically, or look better," Pitts-Taylor, *Surgery Junkies*, 28.

61. Wendy Brown, *Edgework: Critical Essays on Knowledge and Politics* (Princeton: Princeton University Press, 2005), 37, emphasis added.

62. Brown, *Edgework*, 37.

63. Brown, *Edgework*, 42.

64. Brown, *Edgework*, 3.

65. Pitts-Taylor, *Surgery Junkies*, 30.

66. Pitts-Taylor, *Surgery Junkies*, 30.

67. See Jones, *Skintight*; Zylinska, *Bioethics*; Zylinska, "'The Future . . . Is Monstrous': Prosthetics as Ethics," in *The Cyborg Experiments: the Extensions of the Body in the Media Age*, ed. Joanna Zylinska (London: Continuum, 2002).

68. Quoted in Fred Botting and Scott Wilson, "Morlan," in *The Cyborg Experiments: The Extensions of the Body in the Media Age*, ed. Joanna Zylinska (London: Continuum, 2002), 153.

69. Botting and Wilson, "Morlan," 155–156.

70. Botting and Wilson, "Morlan," 155–157.

71. See Rachel Armstrong, "Art, Anger and Medicine: Working with Orlan," in Zylinska, *The Cyborg Experiments*, 172–178.

72. Zylinska, *Bioethics*, 112.

73. See Zylinska, *Bioethics*, 120–123.

74. Jones, *Skintight*, 173.

75. Jones, *Skintight*, 173.

76. See Simon E. Brill, Alex Clarke, David M. Vealeb, and Peter E. M. Butler, "Psychological Management and Body Image Issues in Facial Transplantation," *Body Image* 3, no. 1 (2006), 1–15.

77. See Royal College of Surgeons of England, *Facial Transplantation, Working Party Report*, 2003.

78. *Face/Off* (1997), directed by John Woo.

79. Alex Clarke and Peter E. M. Butler, "Facial Transplantation: Adding to the Reconstructive Options after Severe Facial Injury and Disease," *Expert Opinion on Biological Therapy* 5, no. 12 (2005): 1539–1546, 1539.

80. Royal College of Surgeons of England, *Facial Transplantation*, 1.

81. Royal College of Surgeons of England, *Facial Transplantation*, 1.

82. Royal College of Surgeons of England, *Facial Transplantation*, 20.

83. Clarke and Butler, "Facial Transplantation," 1541.

84. Clarke and Butler, "Facial Transplantation," 1541.

85. Clarke and Butler, "Facial Transplantation," 1542.

86. See, for example, press reporting of the first US full face transplant. The patient, Dallas Weins, "will not resemble 'either what he used to be or the donor' but something in between, said plastic surgeon Dr. Bohdan Pomahac. The 'tissues are really molded on a new person.'" Marilynn Marchione and Russell Contreras, "Texas Man Gets First Full Face Transplant in US," *Associated Press* (March 21, 2011), accessed August 8, 2011, http://www.msnbc.msn.com/id/42192670/ns/health-health_care/.

87. Butler: "this isn't a good example . . . because it's female to male. We'd always do male to male or female to female," in Simon Hattenstone, "Face Saver," *The Guardian Weekend*, November 10 (2007): 62.

88. Hattenstone, "Face Saver," 62.

89. This broadly gave approval for face transplants to proceed in the United Kingdom.

90. Alex Clarke and Peter E. M. Butler, "The Psychological Management of Facial Transplantation," *Expert Review of Neurotherapeutics* 9, no. 7 (2009), 3.

91. John H. Barker, Allen Furr, Michael Cunningham, Federico Grossi, Dalibor Vasilic, Barckley Storey, et al., "Investigation of Risk Acceptance in Facial Transplantation," *Plastic and Reconstructive Surgery*, 118, no. 3 (September 2006), 664.

92. Barker et al., "Investigation of Risk Acceptance," 664.

93. Clarke and Butler, "The Psychological Management," 4.

94. Clarke and Butler, "The Psychological Management," 4.

95. The selection process is understood to be complete at the time of writing. The UK team are awaiting suitable donors.

96. Interview with Alex Clarke, July 14, 2009.

97. Clarke and Butler, "The Psychological Management," 7.

98. Clarke and Butler, "The Psychological Management," 8.

99. Clarke and Butler, "The Psychological Management," 8.

100. Clarke and Butler, "The Psychological Management," 8.

101. Clarke and Butler, "The Psychological Management," 7.

102. Clarke and Butler, "The Psychological Management," 8.

103. Brill et al., "Psychological Management," 6.

104. Brill et al., "Psychological Management," 4.

105. See Nikolas Rose, Pat O'Malley, and Mariana Valverde, "Governmentality," *Annual Review of Law and Social Science* 2 (2006): 83–104.

106. Brill et al., "Psychological Management," 13.

107. Brill et al., "Psychological Management," 3.

108. Brian Masumi, quoted in Mike Featherstone, "Body, Image and Affect in Consumer Culture," *Body and Society* 16, no. 1 (2010), 209.

109. Featherstone, "Body, Image and Affect," 205.

110. Full integration, including especially in a social context, is understood to take five to ten years (interview with Alex Clarke, July 14, 2009).

111. Brill et al., "Psychological Management," 5.

112. Braidotti, *Transpositions*, 148.

113. S. Hettiaratchy, M. A. Randolph, F. Petit, W. P. Lee, P. E. Butler, "Composite Tissue Allotransplantation—A New Era in Plastic Surgery?," *British Journal of Plastic Surgery* 57, no. 5 (2004), 386.

114. See Aldo A. Rossini, Dale L. Greiner, and John P. Mordes, "Induction of Immunologic Tolerance for Transplantation," *Physiological Reviews* 79, no. 1 (1999), 99–129.

115. Hettiaratchy et al., "Composite Tissue Allotransplantation," 388, emphasis added.

116. Hettiaratchy et al., "Composite Tissue Allotransplantation," 389.

117. Isabelle Dinoire received a bone marrow transplant and an injection of donor stem cells, but she remains on a full regime of immunosuppressant drugs. The UK team will also carry out bone marrow transplants after the surgeries have taken place, but, as with Dinoire, this approach will be regarded as "backup only" (interview with Alex Clarke, July 14, 2009).

118. Hettiaratchy et al., "Composite Tissue Allotransplantation," 390.

119. The surgeon was Jean-Michel Dubernard.

120. J. Kanitakis, D. Jullien, P. Petruzzo, N. Hakim, A. Claudy, J. P. Revillard, et al., "Clinicopathologic Features of Graft Rejection of the First Human Hand Allograft," *Transplantation* 76 (2003): 688.

121. See Hettiaratchy et al., "Composite Tissue Allotransplantation."

122. See Brill et al., "Psychological Management."

123. Interview with Alex Clarke, July 14, 2009.

124. See Brill et al., "Psychological Management."

125. Quoted in Hattenstone, "Face Saver," 62.

126. Interview with Alex Clarke, July 14, 2009.

127. Barad, *Meeting the Universe Halfway*, 179.

128. Braidotti, *Transpositions*, 137.

129. Hettiaratchy et al., "Composite Tissue Allotransplantation," 387.

130. Clare Hemmings, "On Reading *Transpositions*: Responses to Rosi Braidotti's *Transpositions: On Nomadic Ethics*," *Subjectivity* 3, no. 2 (2010), 138.

131. Braidotti, *Transpositions*, 208.

Chapter 6

1. Rosi Braidotti, *Transpositions* (Cambridge: Polity Press, 2006), 6.

2. For a useful summary of recent thinking on media ethics, see Lee Wilkins and Clifford G. Christians (eds.), *The Handbook of Mass Media Ethics* (New York: Routledge, 2009).

3. Matthew Kieran, *Media Ethics: A Philosophical Approach* (Westport: Praeger, 1997), 7.

4. John P. Ferré, "A Short History of Media Ethics in the United States," in Wilkins and Christians, *The Handbook of Mass Media Ethics*, 15.

5. Judith Butler provides the following helpful account of Derrida's notion of justice, a notion that serves as a background, or horizon, for such a nonfoundational ethics:.

> Derrida made clear in his short book on Walter Benjamin, *The Force of Law* (1994), that justice was a concept that was yet to come. This does not mean that we cannot expect instances of justice in this life, and it does not mean that justice will arrive for us only in another life. He was clear that there was no other life. It means only that, as an ideal, it is that toward which we strive, without end. Not to strive for justice because it cannot

be fully realized would be as mistaken as believing that one has already arrived at justice and that the only task is to arm oneself adequately to fortify its regime. The first is a form of nihilism (which he opposed) and the second is dogmatism (which he opposed). Derrida kept us alive to the practice of criticism, understanding that social and political transformation was an incessant project, one that could not be relinquished, one that was coextensive with the becoming of life and the encounter with the Other, one that required a reading of the rules by means of which a polity constitutes itself through exclusion or effacement. How is justice done? What justice do we owe others? And what does it mean to act in the name of justice? These were questions that had to be asked regardless of the consequences, and this meant that they were often questions asked when established authorities wished that they were not.

In "Jacques Derrida," *The London Review of Books* 26, no. 21 (November 4, 2004), 31.

6. Peter Foster, "Zuckerberg's Close-Up," *National Post*, January 7, 2011, FP11.

7. Foster, "Zuckerberg's Close-Up," FP11.

8. D. E. Wittkower, "A Reply to Facebook Critics," in *Facebook and Philosophy*, ed. D. E. Wittkower (Chicago: Open Court, 2010), xxii.

9. Wittkower, "A Reply to Facebook Critics," xxii, emphasis added.

10. Wittkower, "A Reply to Facebook Critics," xxv.

11. Wittkower, "A Reply to Facebook Critics," xxv.

12. Mimi Marinucci, "You Can't Front on Facebook," in Wittkower, *Facebook and Philosophy*, 66.

13. James Grimmelmann, "The Privacy Virus," in Wittkower, *Facebook and Philosophy*.

14. See Elizabeth Losh, "With Friends Like These, Who Needs Enemies?," in Wittkower, *Facebook and Philosophy*.

15. See Craig Condella, "Why Can't We Be Virtual Friends?," in Wittkower, *Facebook and Philosophy*.

16. Mariam Thalos, "Why I Am Not a Friend," in Wittkower, *Facebook and Philosophy*, 84.

17. Wittkower, "A Reply to Facebook Critics," xxviii.

18. See Maurice Hamington, "Care Ethics, Friendship, and Facebook," in Wittkower, *Facebook and Philosophy*, 135–145.

19. Marinucci, "You Can't Front on Facebook," 66.

20. Thalos, "Why I Am Not a Friend," 74–77.

21. Michael V. Butera, "Gatekeeper, Moderator, Synthesizer," in Wittkower, *Facebook and Philosophy*, 205.

22. Condella, "Why Can't We Be Virtual Friends?," 119.

23. See Joanna Zylinska, *The Ethics of Cultural Studies* (London: Continuum, 2005); Joanna Zylinska, *Bioethics in the Age of New* Media (Cambridge, MA: MIT Press, 2009); Sarah Kember, "No Humans Allowed? The Alien in/as Feminist Theory," *Feminist Theory* 12, no. 2 (2011): 183–199.

24. See Homero Gil de Zúñiga and Sebastián Valenzuela, "Who Uses Facebook and Why?," in Wittkower, *Facebook and Philosophy*, xxxi–xxxvii.

25. Wittkower, "A Reply to Facebook Critics," xxix.

26. Ian Bogost, "Ian Became a Fan of Marshall McLuhan on Facebook and Suggested You Become a Fan Too," in Wittkower, *Facebook and Philosophy*, 24.

27. Bogost, "Ian Became a Fan of Marshall McLuhan," 25. Graham Meikle is another contributor to *Facebook and Philosophy* ("It's Like Talking to a Wall," 13–20) who takes into account the

"media" aspect of Facebook, not just its "human" side. Meikle analyzes Facebook in terms of media convergence, arguing that it challenges the distinction between private and public, and between the one-to-one communication model and the broadcast model.

28. Bogost, "Ian Became a Fan of Marshall McLuhan," 28.

29. Humberto R. Maturana and Francisco J. Varela, *The Tree of Knowledge: The Biological Roots of Human Understanding*, trans. Robert Paolucci (Boston: Shambhala, 1992); see also Zylinska, *Bioethics*, 39–41.

30. Bogost, "Ian Became a Fan of Marshall McLuhan," 28.

31. Henri Bergson, *Creative Evolution*, trans. Arthur Mitchell (New York: The Modern Library, 1944), 289.

32. Trebor Scholz, "Becoming Facebook," 243–244.

33. See, for example, Melinda Cooper, Andrew Goffey, and Anna Munster (eds.), *Culture Machine*: *Biopolitics* 7 (2005); Vandana Shiva and Ingunn Moser (eds.), *Biopolitics: A Feminist and Ecological Reader on Biotechnology* (London: Third World Network and Zed Books, 1995); Beatriz da Costa and Kavita Philip (eds.), *Tactical Biopolitics: Art, Activism, and Technoscience* (Cambridge, MA: MIT Press, 2008).

34. For an interesting set of critiques of the "lifelessness" of biopolitical theory focusing on the work of Italian philosopher Giorgio Agamben, see "The Agamben Effect," ed. Allison Ross, special issue, *The South Atlantic Quarterly* 107, no. 1 (Winter 2008).

35. Eugene Thacker, *After Life* (Chicago: University of Chicago Press, 2010), xiii.

36. The third conceptualization of life in Western philosophy—one that Thacker describes as "politico-theological," with life seen in terms of spirit (*After Life*, xiii)—necessarily functions as a transcendental horizon against which its more materialist counterpart can be thought and mapped out.

37. N. Katherine Hayles, "Hyper and Deep Attention: The Generational Divide in Cognitive Modes," *Profession* (2007), 187.

38. Hayles, "Hyper and Deep Attention," 187.

39. N. Katherine Hayles, "[-empyre-] post for convergence; print to pixels," posting to the Empyre email list/discussion forum, June, 9, 2010.

40. Hayles, "Hyper and Deep Attention," 192.

41. In a discussion on the Empyre mailing list, "Print to Pixels," in June 2010, Michael Dieter used the term "the neurological turn" to refer to various instances of Internet criticism whereby a direct link is posited between Internet activity and our cognitive abilities and hence our brains. This trope, and everything it represents, is said to be particularly attractive to certain journalistic analyses of new media. As an example of the latter, we can cite: Nicholas Carr, "The Web Shatters Focus, Rewires Brains," *Wired* (June 2010), accessed February 15, 2011, http://m.wired.com/magazine/2010/05/ff_nicholas_carr/all/1.

42. Carr, "The Web Shatters Focus."

43. See, for example, Geert Lovink, "MyBrain.net: The Colonization of Real-Time and Other Trends in Web 2.0," *Eurozine* (March 18, 2010), accessed August 14, 2011, http://www.eurozine.com/articles/2010-03-18-lovink-en.html; Adam Gopnik, "The Information: How the Internet Gets Inside Us," *The New Yorker* (February 14, 2011), accessed August 14, 2011, http://www.newyorker.com/arts/critics/atlarge/2011/02/14/110214crat_atlarge_gopnik.

44. Catherine Malabou, *What Should We Do With Our Brain?*, trans. Sebastian Rand (New York: Fordham University Press, 2003), 3.

45. Malabou, *What Should We Do With Our Brain?*, 11.

46. We are evoking here the term "genohype" as used by Ionat Zurr and Oron Catts in their article "Big Pigs, Small Wings: On Genohype and Artistic Autonomy," in *Culture Machine* 7 (2005). Zurr and Catts write:

> This paper will explore the notion of Genohype, a term coined by Neil Holtzman to describe the discourse of exaggerated claims and overstatements concerning DNA and the Human Genome Project (1999: 409–10). Genohype depicts the hype generated by scientists, the media, the public and the arts with regard to genetic research. In the context of this paper, Genohype is used in relation to the hyperbolic discourse that has been attained by genetic research and its applied outcomes, whether positive or negative. One of the effects of Genohype, as will be illustrated here, is that genetics has become synonymous with all life sciences. (unpaginated)

In a similar vein, "brain-hype" refers to the hyperbolic discourse that has been attained by neuroscience research and its outcomes and to the exaggerated claims that are made with regard to various brain studies.

47. Malabou, *What Should We Do With Our Brain?*, 12.

48. This leads Malabou to argue that "it is no longer possible to distinguish rigorously on an ideological level between 'popularly' accessible neuroscientific studies and the literature of management—including medical management" (*What Should We Do with Our Brain?*, 52).

49. Malabou, *What Should We Do with Our Brain?*, 39.

50. Malabou, *What Should We Do with Our Brain?*, 5.

51. Malabou, *What Should We Do with Our Brain?*, 68.

52. Malabou, *What Should We Do with Our Brain?*, 73.

53. Lovink, "MyBrain.net."

54. Lovink, "MyBrain.net."

55. Bogost, "Ian Became a Fan of Marshall McLuhan," 28.

56. Quoted in Nick Bilton, "The Defense of Computers, the Internet and Our Brains," *New York Times* (June 11, 2010), accessed August 14, 2011, http://bits.blogs.nytimes.com/2010/06/11/in-defense-of-computers-the-internet-and-our-brains.

57. Quoted in Bilton, "The Defense of Computers."

58. Bernard Stiegler, *Taking Care of Youth and the Generations*, trans. Stephen Barker (Stanford: Stanford University Press, 2010), 7.

59. The notion of bio/ethics proposed by us here remains in conversation with Zylinska's earlier work (*Bioethics*). However, there is more emphasis on relationality and mutual co-constitution of media and mediation, and of "us" and "the world" here, and more exploration of *difference as immanence, or differentiation-from-within*, in our book. Yet the relationality that we read through the work of Bergson and Barad in *Life after New Media* remains akin to Zylinska's "ethics of life" as outlined in her *Bioethics*, which was rooted more explicitly in the philosophy of alterity as derived from the thought of Levinas, Derrida, and Stiegler. Any radical distinction between immanence and transcendence, between differentiation-from-within and radical alterity, of course always remains subject to a deconstruction. It is around the notion of "the cut" as already introduced in her *Bioethics* (whereby Zylinska's bioethics considers "the place of

negativity, lack, and . . . 'the cut' in the flow of life as constitutive to it" [31]), and as developed here with regard to Kember's work on duration, entanglement, and relationality ("No Humans Allowed?"), that the two philosophical traditions meet. The slash ("/") in bio/ethics that we have introduced in this chapter foregrounds the decisive role of the notion of "the cut" in this ethical framework.

60. See Thacker, *After Life*.

61. Foster, "Zuckerberg's Close-Up," FP11.

62. Foster, "Zuckerberg's Close-Up," FP11.

63. Bernard Stiegler, *For a New Critique of Political Economy*, trans. Daniel Ross (Cambridge: Polity, 2010), 3.

64. Stiegler, *For a New Critique*, 34.

65. Stiegler, *For a New Critique*, 40.

66. Stiegler, *For a New Critique*, 6.

67. Stiegler, *For a New Critique*, 41.

68. Stiegler, *For a New Critique*, 43.

69. Stiegler, *For a New Critique*, 44.

70. Stiegler, *For a New Critique*, 44.

71. Stanley Cohen, *Folk Devils and Moral Panics: The Creation of Mods and Rockers*, 3rd ed. (London: Routledge, 2002), 3.

72. Stiegler, *For a New Critique*, 42.

73. Stiegler, *For a New Critique*, 10.

74. This is an adaptation of Samuel Beckett's famous line, "All of old. Nothing else ever. Ever tried. Ever failed. No matter. Try again. Fail again. Fail better." From his "Worstward Ho" in *Nohow On* (New York: The Limited Editions Club, 1989). We draw on it further in our conclusion.

75. Malabou, *What Should We Do with Our Brain?*, 36.

76. Stiegler, *For a New Critique*, 104.

77. Malabou, *What Should We Do with Our Brain?*, 62–63, emphasis in original; see also 82.

78. Karen Barad, *Meeting the Universe Halfway* (Durham, NC: Duke University Press, 2007), 142.

79. Andrew Keen, "Your Life Torn Open, Essay 1: Sharing Is a Trap," *Wired.co.uk* (February 3, 2011), accessed August 14, 2011, http://www.wired.co.uk/magazine/archive/2011/03/features/sharing-is-a-trap. The other quotes in that paragraph also from Keen.

80. David Leigh, "You Can Hear Bradley Manning Coming because of the Chains," *The Guardian*, March 16, 2011, accessed March 28, 2011, http://www.guardian.co.uk/media/2011/mar/16/hear-bradley-manning-because-chains.

81. Malabou, *What Should We Do with Our Brain?*, 41.

82. Oliver Burkeman, "Reality Check," *The Guardian*, March 15, 2011, 6–9, 7.

83. Quoted in Burkeman, "Reality Check," 7.

84. Bernard Stiegler, "Technics of Decision: An Interview with Peter Hallward," trans. Sean Gaston, *Angelaki* 8, no. 2 (2003): 164.

85. Federica Frabetti, "Rethinking the Digital Humanities in the Context of Originary Technicity," *Culture Machine* 12 (2011), 16.

86. See the entry in Shaviro's blog ("The Pinocchio Theory") entitled "Fruit Flies and Slime Molds" (December 25, 2010), accessed August 24, 2011, http://www.shaviro.com/Blog/?p=955.

87. In "Fruit Flies and Slime Molds," Shaviro defines a decision as a spontaneously generated activity—as opposed to a selection from a preselected list of options.

88. This argument ties in with the point we made in chapter 4 that "if it is not to be a banal form of vital continuism, the unfolding of life and the emergence of agents in and through relationality requires a 'cut': a decisive in-cision that is also a de-cision, a meaningful yet temporary resolution within mediation, performed from the midst of it."

89. Stiegler, *Taking Care of Youth*, 104.

Chapter 7

An early incarnation of this chapter titled "Creative Media: Performance, Invention, Critique," was published in the book *Interfaces of Performance* (2009), edited by Maria Chatzichristodoulou, Janis Jefferies, and Rachel Zerihan for Ashgate's *Digital Research in the Arts and Humanities* book series (2009). Some of the ideas subsequently went into our follow-up piece, "Creative Media Between Invention and Critique, or What's Still at Stake in Performativity?," which formed an introduction to the special issue of *Culture Machine* that we coedited titled "Creative Media" (vol. 11, 2010).

1. Our use of the terms "progressive" and "conservative" brings together their political connotations that signal, respectively, the overcoming of tradition and its preservation and their biological sense of, respectively, generating new life forms or preserving old ones.

2. John Hartley, "Creative Industries" in *Creative Industries*, ed. John Hartley (Oxford: Blackwell Publishing, 2005), 5.

3. Hartley, "Creative Industries," 5.

4. Andrew Ross, "Nice Work If You Can Get It: The Mercurial Career of Creative Industries Policy," in *MyCreativity Reader: A Critique of Creative Industries*, ed. Geert Lovink and Ned Rossiter (Institute of Network Cultures: Amsterdam, 2007), 19. Ross provides the following rather chilling explanation for this absorption of "creativity" into British economy and policy:

> By the 1990s, the nation's economy was no longer driven by high-volume manufacturing, fuelled by the extractive resources that Bevan had extolled. Like their competitors, Britain's managers were on the lookout for service industries that would "add value" in a distinctive way. In the bowels of Whitehall, an ambitious civil servant came up with a useful statistic. If you lumped all the economic activities of arts and culture professionals together with those in software to create a sector known as the "creative industries," you would have, on paper at least, a revenue powerhouse that generated £60 billion a year (in 2000, revised and improved estimates put the figure at £112 billion). Even more illustrative, the sector was growing at twice the rate of growth of the economy as a whole. For an incoming government, looking to make its mark on the sclerotic post-Thatcher scene, the recent performance and future potential of the creative industries were a godsend. Britain could have its hot new self-image, and Blair's ministers would have the GNP numbers to back it up. Unlike Bevan's coal and fish, or Thatcher's North Sea oil, creativity was a renewable energy resource, mostly untapped; every citizen had some of it, the cost of extraction was minimal, and it would never run out. (23)

5. Ross, "Nice Work If You Can Get It," 27.

6. See Angela McRobbie, *In the Culture Society* (Routledge: London, 1999); David Hesmondhalgh, *The Cultural Industries*, 2nd ed. (London: Sage, 2007); Geert Lovink and Ned Rossiter (eds.), *MyCreativity Reader: A Critique of Creative Industries* (Institute of Network Cultures: Amsterdam, 2007).

7. See Michael Hardt, "Foreword: What Affects are Good For?," in *The Affective Turn: Theorizing the Social*, ed. Patricia Ticineto Clough with Jean Halley (Durham, NC: Duke University Press, 2007).

8. Geert Lovink and Ned Rossiter, "Proposals for Creative Research: Introduction to the *MyCreativity Reader*," in Lovink and Rossiter, *MyCreativity Reader*, 14.

9. Lovink and Rossiter, "Proposals for Creative Research," 14.

10. Lovink and Rossiter, "Proposals for Creative Research," 14.

11. We should also acknowledge here that "the Romantic genius" was in fact a political rather than just aesthetic position, though one that has proved rather double-edged in the American legal system. As explained by Andrew Ross, the "crusade to elevate the authors' labor from that of craftsperson to originator of special value was the heady product of Romantic ideology about the singularity of artistic creation. A multifaceted response to the onset of industrialization and commerce in culture, this ideology was expediently taken up to justify the generous rights extended to authors under copyright law" (*Nice Work If You Can Get It: Life and Labor in Precarious Times* [New York: New York University Press, 2009], 179). The "individual genius" was thus originally positioned as requiring special legal protection precisely due to its individualism and originality, even if in the United States those characteristics were later extended to whole corporations and thus started being used against original "singular" authors.

12. See Ross, *Nice Work If You Can Get It*; McRobbie, *In the Culture Society*; Melissa Gregg, *Work's Intimacy* (Cambridge: Polity, 2011).

13. Angela McRobbie, "Clubs to Companies," in Hartley, *Creative Industries*, 384.

14. David Garcia and Geert Lovink "The ABC of Tactical Media," Tactical Media Network (1997), accessed August 21, 2011, http://video.rutgers.edu/elahi/students/fall06/electronic/garcia_lovnik.pdf.

15. Garcia and Lovink, "The ABC of Tactical Media."

16. Rita Raley, *Tactical Media* (Minneapolis: University of Minnesota Press, 2009), 6.

17. Caroline Bassett, "Cultural Studies and New Media," in *New Cultural Studies: Adventures in Theory*, ed. Gary Hall and Clare Birchall (Edinburgh: Edinburgh University Press, 2006), 233–234.

18. Gilles Deleuze, *Negotiations 1972–1990* (New York: Columbia University Press, 1997), 26–27.

19. Deleuze, *Negotiations*, 27.

20. Deleuze, *Negotiations*, 27–28.

21. Henri Bergson, *Creative Evolution*, trans. Arthur Mitchell (New York: The Modern Library, 1944), 59.

22. Gilles Deleuze, *Essays Critical and Clinical* (London: Verso, 1998), 123.

23. Roberto Esposito, *Bíos: Biopolitics and Philosophy*, trans. Timothy Campbell (Minneapolis: University of Minnesota Press), 81.

24. Esposito, *Bíos*, 89.

25. Esposito, *Bíos*, 95.

26. Quoted in Jeremy Gilbert, "Deleuzian Politics? A Survey and Some Suggestions," *New Formations* 68, no. 1 (2009): 12. Gilbert argues that Protevi's hypothesis "leaves us, for example, with a relatively straightforward answer to the question of whether or not to prefer 'rhizomes' to 'trees.' From this perspective, the answer would be that rhizomes are only preferable to trees to the extent that they are more likely than trees to promote joyous affect. Of course, this leaves open the question of how one knows joyous affect when one sees it, and the very strong assumptions which Deleuze tends to make in implicitly answering it," 12.

27. Gilbert, "Deleuzian Politics?," 22.

28. Gilles Deleuze and Claire Parnet, *Dialogues II* (London: Continuum, 2002), vii.

29. Gilles Deleuze, *Bergsonism* (New York: Zone Books, 1988), 98.

30. Deleuze, *Essays Critical and Clinical*, 118.

31. Gary Hall, Clare Birchall, and Peter Woodbridge, "Deleuze's 'Postscript on the Societies of Control,"' *Culture Machine* 11 (2010), 42.

32. Brian Massumi, *Parables for the Virtual: Movement, Affect, Sensation* (Durham, NC: Duke University Press, 2002), 12.

33. Massumi, *Parables for the Virtual*, 12.

34. Massumi, *Parables for the Virtual*, 12–13.

35. Incidentally, many scholars in philosophy and in media, communications, and cultural studies used to refer to this process of bringing about reality through thinking and writing about it as "performativity"—until they were told by those who had not evidently read Derrida's *Limited Inc.* or Butler's *Gender Trouble* carefully enough that performativity was all about texts and language, and that they now supposedly needed to "turn to materiality."

36. Massumi, *Parables for the Virtual*, 123.

37. Hall, Birchall, and Woodbridge, "Deleuze's 'Postscript on the Societies of Control,'" 42.

38. Hall, Birchall, and Woodbridge, "Deleuze's 'Postscript on the Societies of Control,'" 43.

39. Hall, Birchall, and Woodbridge, "Deleuze's 'Postscript on the Societies of Control,'" 44.

40. See chapter 4 and, in particular, n72 for more on our take on "the affective turn."

41. Michel Foucault, "What Is Critique?," in *What is Enlightenment? Eighteenth-Century Answers and Twentieth-Century Questions*, ed. James Schmidt (Berkeley: University of California Press, 1996), 356.

42. There are similarities here between Foucault's definition and Kant's understanding of the Enlightenment.

43. See Joanna Zylinska, *Bioethics in the Age of New Media* (Cambridge, MA: MIT Press, 2009), 78.

44. Foucault, "What Is Critique?," 383.

45. Judith Butler argues that "this does not mean that no generalizations are possible or that, indeed, we are mired in particularisms. On the contrary, we tread here in an area of constrained generality, one which broaches the philosophical, but must, if it is to remain critical, remain at a distance from that very achievement" in "What Is Critique? An Essay on Foucault's Virtue," *Transversal*, May 2001, accessed September 14, 2011, http://eipcp.net/transversal/0806/butler/en.

46. Butler, "What Is Critique?"

47. Foucault, "What Is Critique?," 383.

48. Isabell Lorey, "Critique and Category," *Transversal* (May 2008), accessed September 14, 2011, http://eipcp.net/transversal/0806/lorey/en.

49. Hall, Birchall, and Woodbridge, "Deleuze's 'Postscript on the Societies of Control,'" 44–45.

50. Jacques Derrida, "On Forgiveness," *On Cosmopolitanism and Forgiveness* (London: Routledge, 2001): 53–54.

51. Gerald Raunig, "What Is Critique? Suspension and Recomposition in Textual and Social Machines," *Transversal*, August 2008, accessed September 14, 2011, http://eipcp.net/transversal/0808/raunig/en.

52. Quoted in Ewa Plonowska Ziarek, *An Ethics of Dissensus: Postmodernity, Feminism, and the Politics of Radical Democracy* (Stanford: Stanford University Press), 41.

53. Ziarek, *An Ethics of Dissensus*, 41.

54. Ziarek, *An Ethics of Dissensus*, 41.

55. Ziarek, *An Ethics of Dissensus*, 41–42.

56. Karen Barad, *Meeting the Universe Halfway* (Durham, NC: Duke University Press, 2007), ix.

57. Barad, *Meeting the Universe Halfway*, ix.

58. Barad, *Meeting the Universe Halfway*, 23.

59. Barad, *Meeting the Universe Halfway*, 33, emphasis in original.

60. Bergson claims that "analysis operates always on the immobile," in Henri Bergson, *The Introduction to Metaphysics*, trans. T. E. Hulme (New York: G. P. Putnam's Sons, 1912), 47.

61. Bergson, *The Introduction to Metaphysics*, 47.

62. This is not an argument for or about human superiority but only about human singularity.

63. This is a quote from Gilles Deleuze and Félix Guattari, *What Is Philosophy?*, trans. Hugh Tomlinson and Graham Burchell (New York: Columbia University Press, 1994), 127.

64. Gilles Deleuze, *Negotiations 1972–1990* (New York: Columbia University Press, 1997), 26–27. In his oft-cited essay, "What Is An Author?" (in *Language, Counter-Memory, Practice*, ed. Donald F. Bouchard, trans. Donald F. Bouchard and Sherry Simon [Ithaca: Cornell University Press, 1977]). Michel Foucault posits the author as a "function of discourse," while also stipulating that this "author-function":

> Is not formed spontaneously through the simple attribution of a discourse to an individual. It results from a complex operation whose purpose is to construct the rational entity we call an author. Undoubtedly, this construction is assigned a "realistic" dimension as we speak of an individual's "profundity" or "creative" power, his intentions or the original inspiration manifested in writing. Nevertheless, these aspects of an individual, which we designate as an author (or which comprise an individual as an author), are projections, in terms always more or less psychological, of our way of handling texts: in the comparisons we make, the traits we extract as pertinent, the continuities we assign, or the exclusions we practice. In addition, all these operations vary according to the period and the form of discourse concerned. (127–128)

65. Gilles Deleuze and Félix Guattari, *A Thousand Plateaus: Capitalism and Schizophrenia*, trans. Brian Massumi (Minneapolis: University of Minnesota Press, 1987), 3.

66. See J. L. Austin, *How to Do Things with Words* (Cambridge, MA: Harvard University Press, 1962); Judith Butler, *Gender Trouble* (New York and London: Routledge, 1990); Judith Butler, *Excitable Speech* (New York and London: Routledge, 1997); and Jacques Derrida, *Limited Inc.* (Evanston, IL: Northwestern University Press, 1988).

67. See Christopher Langton, "Artificial Life," in *The Philosophy of Artificial Life*, ed. Margaret Boden (Oxford: Oxford University Press, 1996).

68. Donna Haraway, *Simians Cyborgs and Women: The Reinvention of Nature* (London: Free Association Books, 1991), 8.

69. Rosi Braidotti, *Metamorphoses* (Cambridge: Polity, 2002), 9.

70. Braidotti, *Metamorphoses*, 9.

71. This is a quote from Gilles Deleuze and Félix Guattari, *What Is Philosophy?* (trans. Hugh Tomlinson and Graham Burchell [New York: Columbia University Press, 1994], 67). They are talking about what they call "half" philosophers who are at the same time "much more than philosophers." Among them Deleuze and Guattari count "Hőlderlin, Kleist, Rimbaud, Mallarmé, Kafka, Michaux, Pessoa, Artaud, and many English and American novelists, from Melville to Lawrence or Miller, in which the reader discovers admiringly that they have written the novel of Spinozism" (67). What is unique about all these writers is that "they do not produce a synthesis of art and philosophy. They branch out and do not stop branching out. They are hybrid geniuses who neither erase nor cover over differences in kind but, on the contrary, use all the resources of their 'athleticism' to install themselves within this very difference, like acrobats torn apart in a perpetual show of strength" (67).

72. This was brought home to Kember in a way that felt both companionable and inspiring when she listened to Katie Mitchell discussing her work *Some Trace of Her* and when she watched—or, rather, attended—the performance itself.

73. The effects of lightning on the body are many and varied. They range from things that you might expect, such as burns, concussion, and heart failure, to things that you probably would not expect, such as neurological damage and changes in personality. One of the ways that lightning can kill you prosaically (and it rarely does) is by means of a ground strike: the negatively charged lower portion of a passing cloud creates a positive charge in the ground underneath it, and this runs up one leg, through your body, and down the other leg. The effect on cattle can be particularly devastating because they have more legs. One of the ways in which lightning can alter you—physically, psychologically, and rather mysteriously—is by using you, instead of a tree, to form an upward streamer: again, the current passes up through you from the ground and connects with the lightning strike descending from above. Sometimes reality is indeed stranger than fiction. When Kember talked about lightning at a recent conference in which she was presenting and performing various aspects of her story, she encountered, for the first time, one of the nonintentional side effects of pursuing an "impersonal mode"—a certain confusion of attendance. Was the science made up and the story really true? Ambiguity is one thing—and it seems totally appropriate to a subject such as cloning. Confusion, however, is perhaps a more problematic if productive affect to manage in the newly recombined writer/reader or speaker/audience relationship.

74. Put bluntly, liberal-humanist, judgmental, moralizing, and conservative.

75. Keith Ansell Pearson, *Philosophy and the Adventure of the Virtual* (London: Routledge, 2002), 8.

76. See Bernard Stiegler, *Technics and Time, 1* (Stanford: Stanford University Press, 1998).

77. Gary Hall and Joanna Zylinska, "Probings: An Interview with Stelarc," in *The Cyborg Experiments*, ed. Joanna Zylinska (London: Continuum, 2002), 114.

78. Hall and Zylinska, "Probings," 115.

79. Geoffrey Batchen, "Electricity Made Visible," in *New Media, Old Media*, ed. Wendy Hui Kyong Chun and Thomas Keenan (New York: Routledge, 2006), 28.

80. For an interesting reading of Fredkin's theories in the context of arts and humanities research see N. Katherine Hayles, *My Mother Was a Computer* (Chicago: University of Chicago Press, 2005).

81. For more on this point, see Gary Hall and Joanna Zylinska, "Experiments of the Stelarc Machine," essay and interview in the catalog accompanying the exhibition *The Obsolete Body*, Centre des Arts, Enghien-les-Bains, France (April 2009).

82. Butler, "What Is Critique?"

83. For more on this issue of body-passivity as an ethical predisposition, see chapter 6 in Joanna Zylinska, *Bioethics in the Age of New Media* (Cambridge, MA: MIT Press, 2009).

Conclusion

This conclusion is largely based on our earlier piece, "Creative Media Between Invention and Critique, or What's Still at Stake in Performativity?," which formed an introduction to the special issue of *Culture Machine* we coedited titled "Creative Media" (vol. 11, 2010).

1. Jacquie Orr, *Panic Diaries: A Genealogy of Panic Disorder* (Durham, NC: Duke University Press, 2005), 6.

2. Quoted from the CERN website, accessed on October 4, 2008, http://public.web.cern.ch/public/en/LHC/LHC-en.html.

3. Lisa Gitelman, *Always Already New: Media, History, and the Data of Culture* (Cambridge, MA: MIT Press), 5.

4. Judith Butler, "What Is Critique? An Essay on Foucault's Virtue," *Transversal*, May 2001, accessed September 14, 2011, http://eipcp.net/transversal/0806/butler/en.

5. In "What Is Critique? Suspension and Recomposition in Textual and Social Machines" (*Transversal*, August 2008, accessed September 14, 2011, http://eipcp.net/transversal/0808/raunig/en), Gerald Raunig argues that critique in the Kantian sense:

> remains an *ars iudicandi*, a technique of distinguishing. . . . All of these revisions of the existing original material are to be understood as a productive process of recomposition. Instead of introducing the distinction as an essentialist excavation of an origin, it is instead a matter of reinstituting a heterogenetic process: not a pure tree schema, at the head of which there is an original text and an *auctor*, but rather a much more winding practice of continual recombination. . . . Critique is thus to be understood as an interplay between the suspended *iudicium* and *inventio*, between the capacity for judgment, which in "making understandable" clearly goes beyond the practice of empirically distinguishing in the sense of separation and exclusion, and the talent for invention that newly concatenates the (significant) components.

6. As some readers can no doubt recognize, the headings that summarize our manifesto are indebted to Samuel Beckett's famous line: "All of old. Nothing else ever. Ever tried. Ever failed. No matter. Try again. Fail again. Fail better." From "Worstward Ho," in *Nohow On* (New York: The Limited Editions Club, 1989).

7. It matters less to us that Baudrillard's answer to the question about events and mediation differs from our own. What matters to us is the question itself.

8. E. M. Forster, *Howards End* (London: Edward Arnold, 1910).

9. We are grateful to Darin Barney for pointing us toward this exhortation.

References

Aarts, Emile, Jan Korst, and Wim F. J. Verhaegh. 2004. Algorithms in Ambient Intelligence. In *Algorithms in Ambient Intelligence*, ed. Wim F. J. Verhaegh, Emile Aarts, and Jan Korst, 1–19. Boston: Kluwer Academic Publishers.

Adam, Alison. 1998. *Artificial Knowing: Gender and the Thinking Machine*. London: Routledge.

Adams, Douglas. 1981. *The Hitch-Hiker's Guide to the Galaxy*. London: Pan Books Ltd.

Adams, Parveen. 1996. *The Emptiness of the Image: Psychoanalysis and Sexual Difference*. London, New York: Routledge.

Adams, Richard. 2010. Selling Fire Insurance to Arsonists. *Guardian* (April 20), accessed July 7, 2011, http://www.guardian.co.uk/commentisfree/cifamerica/2010/apr/20/goldman-sachs-sec-fraud.

Aldiss, Brian. 2001. *Supertoys Last All Summer Long and Other Stories of Future Time*. London: Orbit.

Allon, Fiona. 2003. An Ontology of Everyday Control:Space, Media Flows and "Smart" Living in the Absolute Present. In *Media Space: Place, Scale and Culture in a Media Age*, ed. Nick Couldry and Anna McCarthy, 253–274. London: Routledge.

Andrejevic, Mark. 2007. *iSpy: Surveillance and Power in the Interactive Era*. Lawrence: University Press of Kansas.

Armstrong, Rachel. 2002. Art, Anger, and Medicine: Working with Orlan. In *The Cyborg Experiments: The Extensions of the Body in the Media Age*, ed. Joanna Zylinska, 172–178. London: Continuum.

Austin, J. L. 1962. *How to Do Things with Words*. Cambridge, MA: Harvard University Press.

Barad, Karen. 2007. *Meeting the Universe Halfway*. Durham, NC: Duke University Press.

Barker, John H., A. Furr, M. Cunningham, F. Grossi, D. Vasilic, B. Storey, et al. 2006. "Investigation of Risk Acceptance in Facial Transplantation," *Plastic and Reconstructive Surgery* 118 (3): 663–670.

Barney, Darin. 2007. *One Nation under Google: Citizenship in the Technological Republic*. The 2007 Hart House Lecture. Toronto: Hart House Lectures/Coach House Books.

Barney, Darin. 2007. Radical Citizenship in the Republic of Technology: A Sketch. In *Radical Democracy and the Internet*, ed. Lincoln Dahlberg and Eugenia Siapera, 37–54. New York: Palgrave Macmillan.

Barison, David, and Daniel Ross (dir.). 2004. *The Ister.*

Bassett, Caroline. 2006. Cultural Studies and New Media. In *New Cultural Studies: Adventures in Theory*, ed. Gary Hall and Clare Birchall 220–237. Edinburgh: Edinburgh University Press.

Batchen, Geoffrey. 2006. Electricity Made Visible. In *New Media, Old Media: A History and Theory Reader*, ed. Hui Kyong Chun Wendy and Thomas Keenan, 27–44. New York: Routledge.

Bate, David. 2009. *Photography: Key Concepts*. Oxford: Berg.

Baudrillard, Jean. 1983. *Simulations*, trans. Paul Foss, Paul Patton, and Philip Beitchman. Los Angeles: Semiotext(e).

Baudrillard, Jean. 2002. *The Transparency of Evil: Essays on Extreme Phenomena.* Trans. James Benedict. London: Verso.

Baudrillard, Jean. 2007. *Forget Foucault*, trans. Nicole Dufresne. Los Angeles: Semiotext(e).

Baudrillard, Jean. 2009. *The Gulf War Did Not Take Place*. Sydney: Power Publications.

BBC News. 2007. Northern Rock Gets Bank Bail Out (September 13), accessed July 7, 2011, http://news.bbc.co.uk/1/hi/business/6994099.stm.

Beckett, Samuel. 1989. "Worstward Ho." *In Nohow On*. New York: The Limited Editions Club.

Bergson, Henri. 1912. *The Introduction to Metaphysics*. Trans. T. E. Hulme. New York: G. P. Putnam's Sons.

Bergson, Henri. 1944 [1911]. *Creative Evolution*. Trans. Arthur Mitchell. New York: Random House, The Modern Library.

Bergson, Henri. 1992. *The Creative Mind: An Introduction to Metaphysics*. New York: Kensington Publishing.

Blaxter, Mildred. 2009. The Case of the Vanishing Patient? Image and Experience. *Sociology of Health & Illness* 31 (5): 1–17.

Blum, Virginia L. 2003. *Flesh Wounds. The Culture of Cosmetic Surgery*. Berkeley: University of California Press.

Bogost, Ian. 2010. Ian Became a Fan of Marshall McLuhan on Facebook and Suggested You Become a Fan Too. In *Facebook and Philosophy*, ed. D. E. Wittkower. Chicago: Open Court.

Bolter, Jay David, and Richard Grusin. 2002. *Remediation: Understanding New Media*. Cambridge, MA: MIT Press.

Boorstin, Daniel J. 1961. *The Image: A Guide to Pseudo-Events in America*. New York: Harper Colophon.

Botting, Fred, and Scott Wilson. 2002. Morlan. In *The Cyborg Experiments: The Extensions of the Body in the Media Age*, ed. Joanna Zylinska, 149–167. London: Continuum.

Braidotti, Rosi. 1994. *Nomadic Subjects: Embodiment and Sexual Difference in Contemporary Feminist Theory*. New York: Columbia University Press.

Braidotti, Rosi. 2002. *Metamorphoses*. Cambridge: Polity Press.

Braidotti, Rosi. 2006. *Transpositions: On Nomadic Ethics*. Cambridge: Polity Press.

Briere, Danny, and Pat Hurley. 2007. *Smart Homes for Dummies*. Hoboken, NJ: Wiley Publishing, Inc.

Brill, S. E., A. Clarke, D. M. Veale, and P. E. M. Butler. 2006. Psychological Management and Body Image Issues in Facial Transplantation. *Body Image* 3 (1): 1–15.

Brown, Wendy. 2005. *Edgework: Critical Essays on Knowledge and Politics*. Princeton: Princeton University Press.

Brummer, Alex. 2008. *The Crunch. The Scandal of Northern Rock and the Escalating Credit Crisis*. London: Random House Business Books.

Burgin, Victor. 1982. *Looking at Photographs: Thinking Photography*. London: Macmillan.

Burgin, Victor. 2008. *Components of a Practice*. Milan: Skira.

Burnett, Ron. 2004. *How Images Think*. Cambridge, MA: MIT Press.

Butera, Michael V. 2010. Gatekeeper, Moderator, Synthesizer. In *Facebook and Philosophy*, ed. D. E. Wittkower, 210–212. Chicago: Open Court.

Butler, Judith. 1990. *Gender Trouble*. New York: Routledge.

Butler, Judith. 1997. *Excitable Speech: A Politics of the Performative*. New York: Routledge.

Butler, Judith. 2001. What Is Critique? An Essay on Foucault's Virtue, *Transversal* (May), http://eipcp.net.

Butler, Judith. 2004. Jacques Derrida. *London Review of Books,* November 4, 31.

Burkeman, Oliver. 2011. Reality Check. *Guardian*, March 15, 6–9.

Callon, Michel. 2007. What Does It Mean to Say That Economics Is Performative? In *Do Economists Make Markets?*, ed. Donald Mackenzie, Fabian Muniesa, and Lucia Siu, 311–357. Princeton: Princeton University Press.

Carr, Nicholas. 2008. Is Google Making Us Stupid? *Atlantic* (July–August), accessed December 10, 2011, http://www.theatlantic.com/magazine/archive/2008/07/is-google-making-us-stupid/6868.

Carr, Nicholas. 2010. *The Shallows: How the Internet is Changing the Way We Think, Read and Remember*. New York: WW Norton & Co.

Carr, Nicholas 2010. The Web Shatters Focus, Rewires Brains. *Wired* (May 24), accessed February 15, 2011, http://m.wired.com/magazine/2010/05/ff_nicholas_carr/all/1.

Cass, John, Lorna Goulden, and Slava Kozlov. 2003. Intimate Media: Emotional Needs and Ambient Intelligence. In *The New Everyday: Views on Ambient Intelligence*, ed. Stefano Marzano and Emile Aarts, 218–223. Rotterdam: 010 Publishers.

Castells, Manuel. 1996. *The Rise of the Network Society*. Oxford: Blackwell.

Chalmers, Matthew. 2009. CERN: The View from Inside (January 27), accessed October 30, 2009, http://physicsworld.com/cws/article/indepth/37461.

Chun, Wendy Hui Kyong. 2006. Introduction: Did Somebody Say New Media? In *New Media, Old Media*, ed. Wendy Hui Kyong Chun and Thomas Keenan, 1–10. New York: Routledge.

Cixous, Hélène, and Catherine Clément. 1986. *The Newly Born Woman*. Trans. Betsy Wing. Manchester, UK: Manchester University Press.

Clarke, Alex, and Peter E. M. Butler. 2005. Facial Transplantation: Adding to the Reconstructive Options after Severe Facial Injury and Disease. *Expert Opinion on Biological Therapy* 5 (12): 1539–1546.

Clarke, Alex, and Peter E. M. Butler. 2009. The Psychological Management of Facial Transplantation. *Expert Review of Neurotherapeutics* 9 (7): 1–14.

Clarke, Arthur C. 2000. *2001: A Space Odyssey*. London: Orbit.

Cohen, Stanley. 2002. *Folk Devils and Moral Panics: The Creation of Mods and Rockers*. 3rd ed. London, New York: Routledge.

Colebrook, Claire. 2010. *Deleuze and the Meaning of Life*. London: Continuum.

Condella, Craig. 2010. Why Can't We Be Virtual Friends? In *Facebook and Philosophy*, ed. D. E. Wittkower, 111–121. Chicago: Open Court.

Cooper, Melinda, Andrew Goffey, and Anna Munster, eds. 2005. *Culture Machine: Biopolitics* special issue, vol. 7.

Couldry, Nick. 2003. *Media Rituals: A Critical Approach*. London: Routledge.

Couldry, Nick. 2008. Mediatization or Mediation? Alternative Understandings of the Emergent Space of Digital Storytelling. *New Media & Society* 10 (3): 373–391.

Couldry, Nick, and Anna McCarthy, eds. 2003. *MediaSpace: Place, Scale and Culture in a Media Age*. New York: Routledge.

Cussins, Charis M. 1998. Ontological Choreography: Agency for Women in an Infertility Clinic. In *Differences in Medicine: Unravelling Practices, Techniques and Bodies*, ed. Marc Berg and Annemarie Mol, 166–201. London: Duke University Press.

da Costa, Beatriz, and Kavita Philip, eds. 2008. *Tactical Biopolitics: Art, Activism, and Technoscience*. Cambridge, MA: MIT Press.

Dayan, Daniel. 2010. Beyond Media Events: Disenchantment, Derailment, Disruption. In *Media Events in a Global Age*, ed. Nick Couldry, Andreas Hepp, and Friedrich Krotz, 23–31. London: Routledge.

Dayan, Daniel, and Elihu Katz. 1992. *Media Events: The Live Broadcasting of History*. Cambridge, MA: Harvard University Press.

Davis, Aeron. 2007. *The Mediation of Power: A Critical Introduction*. London: Routledge.

Deleuze, Gilles. 1986. *Cinema 1: The Movement-Image*. Trans. Hugh Tomlinson and Barbara Habberjam. London: The Athlone Press.

Deleuze, Gilles. 1988. *Bergsonism*. Trans. Hugh Tomlinson and Barbara Habberjam. New York: Zone Books.

Deleuze, Gilles. 1997. *Negotiations 1972–1990*. New York: Columbia University Press.

Deleuze, Gilles. 1998. *Essays Critical and Clinical*. London: Verso.

Deleuze, Gilles, and Félix Guattari. 1987. *A Thousand Plateaus: Capitalism and Schizophrenia*. Trans. Brian Massumi. Minneapolis: University of Minnesota Press.

Deleuze, Gilles, and Félix Guattari. 1994. *What Is Philosophy?* Trans. Hugh Tomlinson and Graham Burchell. New York: Columbia University Press.

Gilles, Deleuze, and Claire Parnet. 2002. *Dialogues II*. London: Continuum.

Derrida, Jacques. 1982. Différance. In *Margins of Philosophy*, trans. Alan Bass, 3–27. Chicago: University of Chicago Press.

Derrida, Jacques. 1988. *Limited Inc*. Evanston, IL: Northwestern University Press.

Derrida, Jacques. 2001. *On Cosmopolitanism and Forgiveness*. London: Routledge.

Derrida, Jacques. 2002. Artifactualities. In *Jacques Derrida and Bernard Stiegler, Echographies of Television*. Trans. Jennifer Bajorek. Cambridge: Polity Press.

Derrida, Jacques, and Bernard Stiegler. 2002. *Echographies of Television*. Trans. Jennifer Bajorek. Cambridge: Polity Press.

De Vries, Hent. 1997. Violence and Testimony: On Sacrificing Sacrifice. In *Violence, Identity and Self-Determination*, ed. Hent De Vries and Samuel Weber, 14–43. Stanford: Stanford University Press.

Dick, Philip K. 2004. *UBIK*. London: Orion.

Doane, Mary Ann. 1990. Information, Crisis, Catastrophe. In *Logics of Television*, ed. Patricia Mellencamp, 222–239. Bloomington: Indiana University Press.

Durkheim, Émile. 1968 [1915]. *The Elementary Forms of the Religious Life*. Trans. Joseph Ward Swain. New York: The Free Press.

Edwards, Paul N. 1996. *The Closed World: Computers and the Politics of Discourse in Cold War America*. Cambridge, MA: MIT Press.

Elliott, Larry. 2011. Analysis: Crunch Time for Germany. *Guardian* (November 16), 15.

Ellis, John. 2009. The Fundamental Physics Behind the LHC. In *The Large Hadron Collider. A Marvel of Technology*, ed. Lyndon Evans, 23–37. Lausanne: EPFL Press.

Esposito, Roberto. 2008. *Bíos: Biopolitics and Philosophy*. Trans. Timothy Campbell. Minneapolis: University of Minnesota Press.

Featherstone, Mike. 2010. Body, Image and Affect in Consumer Culture. *Body & Society* 16 (1): 193–221.

Fenton-O'Creavy, Mark. 2008. Has Robert Peston Caused a Recession? Social Amplification, Performativity, and Risks in Financial Markets. Money & Management blog (October 17), accessed July 7, 2011, http://www.open2.net/blogs/money/index.php.

Ferré, John P. 2009. A Short History of Media Ethics in the United States. In *The Handbook of Mass Media Ethics*, ed. Lee Wilkins and Clifford G. Christians, 15–27. New York: Routledge.

Feuer, Jane. 1983. The Concept of Live Television: Ontology as Ideology. In *Regarding Television*, ed. E. Ann Kaplan, 12–21. Frederick, MD: University Publications of America.

Forster, E. M. 1910. *Howards End*. London: Edward Arnold.

Foster, Ian, and Carl Kesselman, eds. 2004. *The Grid: Blueprint for a New Computing Infrastructure*. San Francisco: Morgan Kaufmann.

Foster, Peter. 2011. Zuckerberg's Close-up. *National Post* (January 7), FP11.

Foucault, Michel. 1977. What Is an Author? Trans. Donald F. Bouchard and Sherry Simon. In *Language, Counter-Memory, Practice*, ed. Donald F. Bouchard. Ithaca: Cornell University Press.

Foucault, Michel. 1988. Technologies of the Self. In *Technologies of the Self*, ed. Luther H. Martin, Huck Gutman, and Patrick H. Hutton, 16–49. London: Tavistock Books.

Foucault, Michel. 1996. What Is Critique? In *What Is Enlightenment? Eighteenth-Century Answers and Twentieth-Century Questions*, ed. James Schmidt, 382–398. Berkeley: University of California Press.

Foucault, Michel. 2007. *Security, Territory, Population: Lectures at the Collège de France 1977–1978*. Basingstoke: Palgrave Macmillan.

Foucault, Michel. 2008. *The Birth of Biopolitics: Lectures at the College de France, 1978–79*. Trans. Graham Burchell. Basingstoke: Palgrave Macmillan.

Frabetti, Federica. 2011. Rethinking the Digital Humanities in the Context of Originary Technicity. *Culture Machine* 12.

Frampton, Hollis. 2009. Digressions on the Photographic Agony. In *On the Camera Arts and Consecutive Matters: The Writings of Hollis Frampton*, ed. Bruce Jenkins, 9–21. Cambridge, MA: MIT Press.

Frampton, Hollis. 2009. Eadweard Muybridge: Fragments of a Tesseract. In *On the Camera Arts and Consecutive Matters: The Writings of Hollis Frampton*, ed. Bruce Jenkins, 22–32. Cambridge, MA: MIT Press.

Frampton, Hollis. 2009. Some Propositions on Photography. In *On the Camera Arts and Consecutive Matters: The Writings of Hollis Frampton*, ed. Bruce Jenkins, 5–8. Cambridge, MA: MIT Press.

Friday, Jonathan. 2006. Stillness Becoming: Reflections on Bazin, Barthes, and Photographic Stillness. In *Stillness and Time: Photography and the Moving Image*, ed. David Green and Joanna Lowry, 39–54. Brighton, UK: Photoworks.

Friedan, Betty. 2010. *The Feminine Mystique, with a new introduction by Lionel Shriver*. London: Penguin Books.

Fuller, Matthew. 2005. *Media Ecologies: Materialist Energies in Art and Technoculture*. Cambridge, MA: MIT Press.

Garcia, David, and Geert Lovink. 1997. The ABC of Tactical Media. Tactical Media Network, accessed August 21, 2011, http://video.rutgers.edu/elahi/students/fall06/electronic/garcia_lovnik .pdf.

Gell, Alfred. 1992. The Technology of Enchantment and the Enchantment of Technology. In *Anthropology, Art and Aesthetics*, ed. Jeremy Coote and Anthony Shelton, 40–66. Oxford: Clarendon.

Gilbert, Jeremy. 2009. Deleuzian Politics? A Survey and Some Suggestions. *New Formations* 68 (1): 10–33.

Gil de Zúñiga, Homero, and Sebastián Valenzuela. 2010. Who Uses Facebook and Why? In *Facebook and Philosophy*, ed. D. E. Wittkower, xxxi–xxxvii. Chicago: Open Court.

Gitelman, Lisa. 2006. *Always Already New: Media, History, and the Data of Culture*. Cambridge, MA: MIT Press.

Gopnik, Adam. 2011. The Information: How the Internet Gets Inside Us. *New Yorker* (February 14), accessed August 14, 2011, http://www.newyorker.com/arts/critics/atlarge/2011/02/14/ 110214crat_atlarge_gopnik.

Greenfield, Adam. 2006. *Everyware: The Dawning Age of Ubiquitous Computing*. Berkeley: New Riders Publishing.

Gregg, Melissa. 2011. *Work's Intimacy*. Cambridge: Polity.

Grimmelmann, James. The Privacy Virus. In *Facebook and Philosophy*, ed. D. E. Wittkower, 1–12. Chicago: Open Court.

Grosz, Elizabeth. 1999. Becoming . . . An Introduction. In *Becomings: Explorations in Time, Memory, and Futures*, ed. Elizabeth Grosz, 1–11. Ithaca: Cornell University Press.

Grosz, Elizabeth. 2004. *The Nick of Time: Politics, Evolution, and the Untimely*. Durham: Duke University Press.

Grusin, Richard. 2010. *Premediation: Affect and Mediality after 9/11*. Basingstoke: Palgrave Macmillan.

Gumpert, Gary, and Robert Cathcart. 1990. A Theory of Mediation. In *Mediation, Information, and Communication*, ed. B. Ruben and L. Lievrouw, 21–36. New Brunswick, NJ: Transaction.

Gray, Chris Hables. 2002. *Cyborg Citizen*. New York: Routledge.

Hacking, Ian. 1983. *Representing and Intervening: Introductory Topics in the Philosophy of Natural Science*. Cambridge: Cambridge University Press.

Hall, Gary. 2008. *Digitize This Book! The Politics of New Media, or Why We Need Open Access Now*. Minneapolis: University of Minnesota Press.

Hall, Gary, Clare Birchall, and Peter Woodbridge. 2010. Deleuze's "Postscript on the Societies of Control." *Culture Machine* 11.

Hall, Gary, and Joanna Zylinska. 2002. Probings: An Interview with Stelarc. In *The Cyborg Experiments*, ed. Joanna Zylinska. London: Continuum.

Hall, Gary, and Joanna Zylinska. 2009. Experiments of the Stelarc Machine. Essay and interview in the catalog accompanying the exhibition *The Obsolete Body* (April). France: Centre des Arts, Enghien-les-Bains.

Hamington, Maurice. 2010. Care Ethics, Friendship, and Facebook. In *Facebook and Philosophy*, ed. D. E. Wittkower, 135–145. Chicago: Open Court.

Hanan, Joshua S. 2010. Home Is Where the Capital Is: The Culture of Real Estate in an Era of Control Societies. *Communication and Critical/Cultural Studies* 7 (2): 176–201.

Hansen, Mark B. N. 2004. *New Philosophy for New Media*. Cambridge, MA: MIT Press.

Haraway, Donna. 1991. *Simians Cyborgs and Women: The Reinvention of Nature*. London: Free Association Books.

Hardt, Michael. 2007. Foreword: What Affects are Good For? In *The Affective Turn: Theorizing the Social*, ed. Patricia Ticineto Clough with Jean Halley, ix–xiii. Durham, NC: Duke University Press.

Hartley, John. 2005. Creative Industries. In *Creative Industries*, ed. John Hartley, 1–40. Oxford: Blackwell Publishing.

Hattenstone, Simon. 2007. Face Saver, *Guardian Weekend* (November 10), 58–66.

Hayles, N. Katherine. 1999. *How We Became Posthuman: Virtual Bodies in Cybernetics, Literature and Informatics*. Chicago: University of Chicago Press.

Hayles, N. Katherine. 2005. *My Mother Was a Computer: Digital Subjects and Literary Texts*. Chicago: University of Chicago Press.

Hayles, N. Katherine. 2007. Hyper and Deep Attention: The Generational Divide in Cognitive Modes. *Profession* 13:187–199.

Hayles, N. Katherine. 2010. [-empyre-] post for convergence; print to pixels. Posting to Empyre email list/discussion forum (June 9).

Heidegger, Martin. 1977. The Question Concerning Technology. In *Martin Heidegger, Basic Writings*, ed. David Farrell Krell, 287–317. New York: Harper & Row.

Helmreich, Stefan. 1998. *Silicon Second Nature: Culturing Artificial Life in a Digital World*. Berkeley: University of California Press.

Hemmings, Clare. 2005. Invoking Affect: Cultural Theory and the Ontological Turn. *Cultural Studies* 19 (5): 548–567.

Hemmings, Clare. 2010. On Reading *Transpositions*: Responses to Rosi Braidotti's *Transpositions: On Nomadic Ethics*. *Subjectivity* 3 (2): 136–140.

Hepp, Andreas, and Nick Couldry. 2010. Introduction: Media Events in Globalized Media Cultures. In *Media Events in a Global Age*, ed. Nick Couldry, Andreas Hepp, and Friedrich Krotz, 1–21. London: Routledge.

Hesmondhalgh, David. 2007. *The Cultural Industries*. 2nd ed. London: Sage.

Hettiaratchy, Shehan, Mark A. Randolph, François Petit, W. P. Andrew Lee, and Peter E. M. Butler. 2004. Composite Tissue Allotransplantation: A New Era in Plastic Surgery? *British Journal of Plastic Surgery* 57 (5): 381–391.

Jones, Meredith. 2008. *Skintight: An Anatomy of Cosmetic Surgery*. Oxford: Berg.

Holland, Eugene. 2005. Concepts and Utopia. In *The Deleuze Dictionary*, ed. Adrian Parr, 52–53. Edinburgh: Edinburgh University Press.

Kanitakis, J., D. Jullien, P. Petruzzo, N. Hakim, A. Claudy, J. P. Revillard, et al. 2003. Clinicopathologic Features of Graft Rejection of the First Human Hand Allograft. *Transplantation* 76:688–693.

Katz, Elihu, and Tamar Liebes. 2010. "NO MORE PEACE!": How Disaster, Terror, and War Have Upstaged Media Events. In *Media Events in a Global Age*, ed. Nick Couldry, Andreas Hepp, and Friedrich Krotz, 32–42. London: Routledge.

Keen, Andrew. 2011. Your Life Torn Open, Essay 1: Sharing Is a Trap. *Wired* (February 3), accessed August 14, 2011, http://www.wired.co.uk/magazine/archive/2011/03/features/sharing-is-a-trap.

Kember, Sarah. 2003. *Cyberfeminism and Artificial Life*. London: Routledge.

Kember, Sarah. 2004. Doing Technoscience as (New) Media. In *Media and Cultural Theory*, ed. James Curran and David Morley, 235–249. London: Routledge.

Kember, Sarah. 2006. Creative Evolution? The Quest for Life (on Mars). *Culture Machine: InterZone* (March).

Kember, Sarah. 2008. The Virtual Life of Photography. *Photographies* 1 (2): 175–203.

Kember, Sarah. 2010. Media, Mars, Metamorphosis. *Culture Machine* 11.

Kember, Sarah. 2011. No Humans Allowed? The Alien in/as Feminist Theory. *Feminist Theory* 12 (2): 183–199.

Kember, S. J. 2011. *The Optical Effects of Lightning*. Newcastle upon Tyne: Wild Wolf Publishing.

Kember, Sarah, and Joanna Zylinska. 2009. Creative Media: Performance, Invention, Critique. In *Interfaces of Performance. Aldergate*, ed. Maria Chatzichristodoulou, Janis Jefferies, and Rachel Zerihan, 7–23. Aldergate, UK: Ashgate.

Kember, Sarah, and Joanna Zylinska. 2010. Creative Media between Invention and Critique, or What's Still at Stake in Performativity? *Culture Machine* 11.

Kieran, Matthew. 1997. *Media Ethics: A Philosophical Approach*. Westport, CT: Praeger.

Kirkpatrick, David. 2010. *The Facebook Effect: The Inside Story of the Company that Is Connecting the World*. New York: Simon & Schuster.

Kubrick, Stanley (dir.). 1968. *2001: A Space Odyssey*.

Langton, Christopher G. 1996. Artificial Life. In *The Philosophy of Artificial Life*, ed. Margaret A. Boden, 39–95. Oxford: Oxford University Press.

Lasch, Christopher. 1979. *The Culture of Narcissism*. London: Abacus.

Lash, Scott, and Celia Lury. 2007. *Global Culture Industry: The Mediation of Things*. Cambridge: Polity.

Leigh, David. 2011. You Can Hear Bradley Manning Coming Because of the Chains. *Guardian* (March 16).

Lem, Stanislaw. 1984 [1967]. *Summa Technologiae*. 4th ed. Lublin, Poland: Wydawnictwo Lubelskie.

Levinas, Emmanuel. 1969. *Totality and Infinity: An Essay on Exteriority*. Trans. Alphonso Lingis. Pittsburgh: Duquesne University Press.

Levinas, Emmanuel. 1986. The Trace of the Other. Trans. Alphonso Lingis. In *Deconstruction in Context*, ed. Mark C. Taylor, 345–359. Chicago: University of Chicago Press.

Lievrouw, Leah A., and Sonia Livingstone, eds. 2006. *The Handbook of New Media*. London,: Sage.

Lister, Martin. 1995. Introductory Essay. In *The Photographic Image in Digital Culture*, ed. Martin Lister, 1–26. London: Routledge.

Lister, Martin, Jon Dovey, Seth Giddings, Iain Grant, and Kieran Kelly. 2008. *New Media: A Critical Introduction*. 2nd ed. New York: Routledge.

Lorey, Isabell. 2008. Critique and Category. *Transversal* (May), accessed September 14, 2011, http://eipcp.net/transversal/0806/lorey/en.

Losh, Elizabeth. 2010. With Friends Like These, Who Needs Enemies? In *Facebook and Philosophy*, ed. D. E. Wittkower. Chicago: Open Court.

Lotringer, Sylvère. 2007. Introduction: Exterminating Angel. In Jean Baudrillard, *Forget Foucault*, 10–13. Los Angeles: Semiotext(e).

Lovink, Geert. 2010. MyBrain.net: The Colonization of Real-Time and Other Trends in Web 2.0. *Eurozine* (March 18), accessed August 14, 2011, http://www.eurozine.com/articles/2010-03-18-lovink-en.html.

Lovink, Geert, and Ned Rossiter. 2007. Proposals for Creative Research: Introduction to the MyCreativity Reader. In *MyCreativity Reader: A Critique of Creative Industries*, ed. Geert Lovink and Ned Rossiter. Amsterdam: Institute of Network Cultures.

Lovink, Geert, and Ned Rossiter, eds. 2007. *MyCreativity Reader: A Critique of Creative Industries*. Amsterdam: Institute of Network Cultures.

Lyon, David. 2003. *Surveillance as Social Sorting*. London: Routledge.

Lyon, David. 2008. *Surveillance after September 11*. Cambridge: Polity.

Mackenzie, Donald, Fabian Muniesa, and Lucia Siu Callon. 2007. Introduction. In *Do Economists Make Markets?* ed. Donald Mackenzie, Fabian Muniesa, and Lucia Siu, 1–19. Princeton: Princeton University Press.

Malabou, Catherine. 2008. *What Should We Do with Our Brain?* Trans. Sebastian Rand. New York: Fordham University Press.

Manovich, Lev. 2001. *The Language of New Media*. Cambridge, MA: MIT Press.

Marchione, Marilynn, and Russell Contreras. 2011. Texas Man Gets First Full Face Transplant in US. *The Associated Press* (March 21), accessed August 8, 2011, http://www.msnbc.msn.com/id/42192670/ns/health-health_care/.

Marinucci, Mimi. 2010. You Can't Front on Facebook. In *Facebook and Philosophy*, ed. D. E. Wittkower. Chicago: Open Court.

Marzano, Stefano. 2003. Cultural Issues in Ambient Intelligence. In *The New Everyday: Views on Ambient Intelligence*, ed. Stefano Marzano and Emile Aarts, 8–11. Rotterdam: 010 Publishers.

Marvin, Carolyn. 1988. *When Old Technologies Were New*. New York: Oxford University Press.

McRobbie, Angela. 1999. *In the Culture Society*. London: Routledge.

McRobbie, Angela. 2005. Clubs to Companies. In *Creative Industries*, ed. John Hartley, 375–390. Oxford: Blackwell Publishing.

McRobbie, Angela. 2009. *The Aftermath of Feminism: Gender, Culture and Social Change*. London: Sage.

Maslow, Abraham. 1987. *Motivation and Personality*. New York: Longman.

Massumi, Brian. 2002. *Parables for the Virtual: Movement, Affect, Sensation*. Durham, NC: Duke University Press.

Maturana, Humberto R., and Francisco J. Varela. 1992. *The Tree of Knowledge: The Biological Roots of Human Understanding*. Trans. Robert Paolucci. Boston: Shambhala.

Meikle, Graham. 2010. It's Like Talking to a Wall. In *Facebook and Philosophy*, ed. D. E. Wittkower. Chicago: Open Court.

Merrin, William. 2005. *Baudrillard and the Media*. Cambridge: Polity Press.

Miller, Frederic P., Agnes F. Vandome, and John McBrewster, eds. 2009. *Monsanto*. Beau Bassin, Mauritius: Alphascript Publishing.

Morley, David. 2000. *Home Territories: Media, Mobility, and Identity*. London: Routledge.

Orr, Jackie. 2005. *Panic Diaries: A Genealogy of Panic Disorder*. Durham, NC: Duke University Press.

Padgham, Lin, and Michael Winikoff. 2004. *Developing Intelligent Agent Systems*. Chichester: John Wiley and Sons Ltd.

Patton, Paul, and John Protevi. 2003. Introduction. In *Between Deleuze and Derrida*, ed. Paul Patton and John Protevi, 1–14. London: Continuum.

Pearson, Keith Ansell. 2002. *Philosophy and the Adventure of the Virtual*. London: Routledge.

Penley, Constance. 1997. *NASA/Trek: Popular Science and Sex in America*. New York: Verso.

Peston, Robert. 2008. The New Capitalism. BBC News (December 8), accessed July 7, 2011, http://www.bbc.co.uk/blogs/thereporters/robertpeston/newcapitalism.pdf.

Peters, Benjamin. 2009. And Lead Us Not into Thinking the New is New: A Bibliographic Case for New Media History. *New Media & Society* 11 (1–2): 13–30.

Picard, Rosalind. 1996. Does HAL Cry Digital Tears? Emotion and Computers. In *Hal's Legacy: 2001's Computer as Dream and Reality*, ed. David G. Stork, 279–305. Cambridge, MA: MIT Press.

Pickering, Andrew. 1995. *The Mangle of Practice: Time, Agency, and Science*. Chicago: The University of Chicago Press.

Pierce, Andrew. 2008. The Queen Asks Why No One Saw the Credit Crunch Coming. *Telegraph* (November 5), accessed July 7, 2011, http://www.telegraph.co.uk/news/uknews/theroyalfamily/3386353/The-Queen-asks-why-no-one-saw-the-credit-crunch-coming.html.

Pitts-Taylor, Victoria. 2007. *Surgery Junkies: Wellness and Pathology in Cosmetic Surgery*. New Brunswick, NJ: Rutgers University Press.

Poster, Mark. 2002. High-Tech Frankenstein, or Heidegger Meets Stelarc. In *The Cyborg Experiments: The Extensions of the Body in the Media Age*, ed. Joanna Zylinska, 15–32. London: Continuum.

Raley, Rita. 2009. *Tactical Media*. Minneapolis: University of Minnesota Press.

Raunig, Gerald. 2008. What Is Critique? Suspension and Recomposition in Textual and Social Machines. *Transversal* (August), accessed September 14, 2011, http://eipcp.net/transversal/0808/raunig/en.

Ritchin, Fred. 1991. The End of Photography as We Have Known It. In *Photovideo: Photography in the Age of the Computer*, ed. Paul Wombell, 8–15. London: Rivers Oram Press.

Rose, Nikolas, Pat O'Malley, and Mariana Valverde. 2006. Governmentality, *Annual Review of Law and Social Science* 2:83–104.

Ross, Alison, ed. 2008. The Agamben Effect, special issue, *The South Atlantic Quarterly* 107 (1).

Ross, Andrew. 2007. Nice Work if You Can Get It: The Mercurial Career of Creative Industries Policy. In *MyCreativity Reader: A Critique of Creative Industries*, ed. Geert Lovink and Ned Rossiter, 19–41. Amsterdam: Institute of Network Cultures.

Ross, Andrew. 2009. *Nice Work If You Can Get It: Life and Labor in Precarious Times*. New York: New York University Press.

Rossini, A. A., D. L. Greiner, and J. P. Mordes. 1999. Induction of Immunologic Tolerance for Transplantation. *Physiological Reviews* 79 (1): 99–129.

Royal College of Surgeons of England. 2003. *Facial Transplantation. Working Party Report*. The Royal College of Surgeons of England.

Scholz, Trebor. 2010. Becoming Facebook. In *Facebook and Philosophy*, ed. D. E. Wittkower. Chicago: Open Court.

Seamark, Michael. 2008. Does This BBC Man Have Too Much Power? Reporter Blamed for Helping Trigger Shares Fall. *Daily Mail* (October 8), accessed July 7, 2011, http://www.dailymail.co.uk/news/article-1072549/BBC-reporter-Robert-Peston-blamed-helping-trigger-shares-fall.html.

Sekatskiy, Alexander. 2010. The Photographic Argument of Philosophy. *Philosophy of Photography* 1 (1): 81–88.

Shaviro, Steven. 2010. Fruit Flies and Slime Molds., The Pinocchio Theory blog (December 25), accessed August 24, 2011, http://www.shaviro.com/Blog/?p=955.

Shils, Edward. 1975. *Center and Periphery: Essays in Macrosociology*. Chicago: University of Chicago Press.

Shils, Edward A., and Michael Young. 1975 [1956]. The Meaning of the Coronation. In *Center and Periphery: Essays in Macrosociology*, ed. Edward Shils. Chicago: University of Chicago Press.

Shirky, Clay. 2008. *Here Comes Everybody: The Power of Organizing without Organizations*. New York: Penguin Press.

Shirky, Clay. 2010. *Cognitive Surplus: Creativity and Generosity in a Connected Age*. New York: Penguin Press.

Shiva, Vandana, and Ingunn Moser, eds. 2005. *Biopolitics: A Feminist and Ecological Reader on Biotechnology*. London: Third World Network and Zed Books.

Smith, Daniel W. 2003. Deleuze and Derrida, Immanence and Transcendence: Two Directions in Recent French Thought. In *Between Deleuze and Derrida*, ed. Paul Patton and John Protevi, 46–66. London: Continuum.

Sontag, Susan. 2004. Regarding the Torture of Others. *New York Times* (May 23), accessed July 14, 2011, http://www.nytimes.com/2004/05/23/magazine/regarding-the-torture-of-others.html.

Stiegler, Bernard. 1998. *Technics and Time, 1: The Fault of Epimetheus*. Trans. Richard Beardsworth and George Collins. Stanford: Stanford University Press.

Stiegler, Bernard. 2003. Technics of Decision: An Interview with Peter Hallward. Trans. Sean Gaston. *Angelaki* 8 (2): 151–168.

Stiegler, Bernard. 2009. *Technics and Time, 2: Disorientation*. Trans. Stephen Barker. Stanford: Stanford University Press.

Stiegler, Bernard. 2010. *For a New Critique of Political Economy*. Trans. Daniel Ross. Cambridge: Polity.

Stiegler, Bernard. 2010. *Taking Care of Youth and the Generations*. Trans. Stephen Barker. Stanford: Stanford University Press.

Stork, David G., ed. 1996. *Hal's Legacy: 2001's Computer as Dream and Reality*. Cambridge, MA: MIT Press.

Suchman, Lucy. 2007. *Human–Machine Reconfigurations*. Cambridge: Cambridge University Press.

Teather, David. 2010. End of Story for Hardbacks? Amazon's Ebook Milestone. *Guardian* (July 21), 1.

Thacker, Eugene. 2003. What Is Biomedia? *Configurations* 11 (1): 47–79.

Thacker, Eugene. 2010. *After Life*. Chicago: University of Chicago Press.

Thalos, Mariam. 2010. Why I Am Not a Friend. In *Facebook and Philosophy*, ed. D. E. Wittkower. Chicago: Open Court.

Turing, Alan. 1950. Computing Machinery and Intelligence. *Mind* 59 (October): 433–460.

Turner, Graham. 2008. *The Credit Crunch*. London: Pluto.

Turow, Joseph. 2008. *Niche Envy*. Cambridge, MA: MIT Press.

Van Loon, Joost. 2007. *Media Technology: Critical Perspectives*. Maidenhead, UK: Open University Press.

Van Loon, Joost. 2010. Modalities of Mediation. In *Media Events in a Global Age*, ed. Nick Couldry, Andreas Hepp, and Friedrich Krotz, 109–123. London: Routledge.

Verhaegh, Wim F. J., Emile Aarts, and Jan Korst, eds. 2004. *Algorithms in Ambient Intelligence*. Boston: Kluwer Academic Publishers.

Wegenstein, Bernadette. 2006. *Getting under the Skin: The Body and Media Theory*. Cambridge, MA: MIT Press.

Weiser, Mark. 1991. The Computer for the 21st Century. *Scientific American* 265 (September): 94–104.

White, Mimi. 2003. The Attractions of Television: Reconsidering Liveness. In *Media Space: Place, Scale, and Culture in a Media Age,* ed. Nick Couldry and Anna McCarthy, 75–91. London: Routledge.

Wilczek, Frank. 2008. *The Lightness of Being: Mass, Ether, and the Unification of Forces.* New York: Basic Books.

Wilkins, Lee, and Clifford G. Christians, eds. 2009. *The Handbook of Mass Media Ethics.* New York: Routledge.

Wilkinson, Carl. 2007. *The Observer Book of Space.* London: Observer Books.

Wittkower, D. E. 2010. A Reply to Facebook Critics. In *Facebook and Philosophy,* ed. D. E. Wittkower, xxi–xxx. Chicago: Open Court.

Ziarek, Ewa Plonowska. 2001. *An Ethics of Dissensus: Postmodernity, Feminism, and the Politics of Radical Democracy.* Stanford: Stanford University Press.

Žižek, Slavoj. 1999. *The Ticklish Subject: The Absent Centre of Political Ontology.* London: Verso.

Zurr, Ionat, and Oron Catts. 2005. Big Pigs, Small Wings: On Genohype and Artistic Autonomy. *Culture Machine* 7.

Zylinska, Joanna. 2002. "The Future . . . Is Monstrous": Prosthetics as Ethics. In *The Cyborg Experiments: The Extensions of the Body in the Media Age,* ed. Joanna Zylinska, 214–236. London: Continuum.

Zylinska, Joanna. 2005. *The Ethics of Cultural Studies.* London: Continuum.

Zylinska, Joanna. 2008. The Cut of the Artist: Sellars' Anatomy Lesson. Catalog essay to accompany Nina Sellars's exhibition at Guildford Lane Gallery, Melbourne.

Zylinska, Joanna. 2009. *Bioethics in the Age of New Media.* Cambridge, MA: MIT Press.

Zylinska, Joanna. 2010. *If It Reads, It Bleeds* (video, 3 min.).

Zylinska, Joanna. 2010. On Bad Archives, Unruly Snappers, and Liquid Photographs. *Photographies* 3 (2): 139–153.

Zylinska, Joanna. 2010. Playing God, Playing Adam: The Politics and Ethics of Enhancement. *Journal of Bioethical Inquiry* 7 (2): 149–161.

Index